PROBABILITY FOR ENGINEERING
with Applications to Reliability

ELECTRICAL ENGINEERING, COMMUNICATIONS, AND SIGNAL PROCESSING

Raymond L. Pickholtz, Series Editor

Computer Network Architectures
Anton Meijer and Paul Peeters

Spread Spectrum Communications, Volume I
Marvin K. Simon, Jim K. Omura, Robert A. Scholtz, and Barry K. Levitt

Spread Spectrum Communications, Volume II
Marvin K. Simon, Jim K. Omura, Robert A. Scholtz, and Barry K. Levitt

Spread Spectrum Communications, Volume III
Marvin K. Simon, Jim K. Omura, Robert A. Scholtz, and Barry K. Levitt

Elements of Digital Satellite Communication: System Alternatives, Analyses and Optimization, Volume I
William W Wu

Elements of Digital Satellite Communication: Channel Coding and Integrated Services Digital Satellite Networks, Volume II
William W Wu

Current Advances in Distributed Computing and Communications
Yechiam Yemini

Digital Transmission Systems and Networks, Volume II: Applications
Michael J. Miller and Syed V. Ahamed

Transmission Analysis in Communication Systems, Volume I
Osamu Shimbo

Transmission Analysis in Communication Systems, Volume II
Osamu Shimbo

Spread Spectrum Signal Design: LPE and AJ Systems
David L. Nicholson

Digital Signal Processing Design
Andrew Bateman and Warren Yates

Probability for Engineering with Applications to Reliability
Lavon B. Page

OTHER WORKS OF INTEREST

Local Area and Multiple Access Networks
Raymond L. Pickholtz, Editor

Telecommunications and the Law: An Anthology
Walter Sapronov, Editor

PROBABILITY FOR ENGINEERING
with Applications to Reliability

Lavon B. Page
North Carolina State University

COMPUTER SCIENCE PRESS

Library of Congress Cataloging-in-Publication Data

Page, Lavon B.
Probability for engineering with applications to reliability
Lavon B. Page.
p. cm.
Bibliography: p.
Includes index.
ISBN 0-7167-8187-5
1. Probabilities. 2. Engineering mathematics. I. Title.
QA273.P19 1988 88-29926
519.2—dc19 CIP

Printed in the United States of America

Computer Science Press

1 2 3 4 5 6 7 8 9 0 RRD 7 6 5 4 3 2 1 0 8 9

for Jo Ellen

Contents

Preface

This book is an attempt to merge the best of two worlds. It brings precise mathematical language and terminology into an arena of modern engineering reliability probems. These include such topics as simple analysis of circuits, design of redundancy into systems, reliability of communication networks, mean time to failure, time-dependent system reliability, and uses of probability in recursive solutions to real-world problems.

All books reflect the point of view of the author. Texts on probability written by engineers usually move as quickly as possible to the applications of most interest to the writer. The mathematical foundations are presented casually and quickly, if at all, and understanding is gleaned from examples rather than from definitions and theorems. Mathematicians, on the other hand, tend to write probability books that dwell on topics such as combinatorics, labeling problems, allocation schemes, and limit theorems. Such books present little that seems relevant to the world of an undergraduate engineering student.

In teaching probability to engineering students for more than a decade, I have seen the difficulties caused by both of these kinds of textbooks. Students with good intuition are often handicapped from never having come to grips with the basic concepts. On the other hand, the importance of fundamental concepts escapes the student unless some application is in sight. For this reason, significant applications appear early in this text. Chapter 2, for example, illustrates how the idea of conditional probability can be blended with algorithmic problem solving to develop tools for reliability analysis of complex systems such as circuits, communication networks, and chemical reactors. These contemporary problems mix probability and discrete mathematics, and they require a solid understanding of the basics. But the student is rewarded with a sense of relevance that isn't matched by drawing balls out of an urn.

The essentials of an introduction to probability are found in Chapters 1, 3, 4, the first three sections of Chapter 5, and the first four sections of Chapter 6. The remaining sections of Chapter 6 introduce conditional density functions, conditional expectation, and the central limit theorem. (The De Moivre-Laplace version of the central limit theorem appears in Chapter 5.) Chapter 7 gives a brief introduction to stochastic processes, with heavy emphasis on the Poisson process. Chapter 2 applies the basic ideas of probability to reliability problems involving a variety of complex systems. Chapter 8 shows how to extend the concept of reliability to systems having components whose reliabilities vary with time.

Proper mathematical language and detail are important in formulating correct mathematical models, and this book reflects that fact. However, the book is practical rather than theoretical, and logical explanations or proofs are given only where they are an aid to intuitive understanding.

The real world confronts us with some easy problems and some hard ones. So does this book. As a result, it does not seem appropriate to treat all problems the same with regard to hints given or answers provided. This has prompted the inclusion of the section of answers, partial solutions, and hints to selected problems that appears as Appendix A. My hope is that students will give serious thought to problems *before* consulting this appendix, and that *after* consulting it they will think seriously about alternative solutions suggested or fill in missing details.

My own research in recent years has shifted into the area of reliability analysis and risk assessment. And, just as with other books, this one reflects the author's own interests. The common theme in this book is the search for reliability in an increasingly complex world. News reports on topics as diverse as arms control, strategic defense initiatives, hazardous waste dumps, and nuclear reactors all remind us of the fact that uncertainty (or lack of reliability) is central to many of the pressing issues of our day.

Lavon B. Page
September 1988

PROBABILITY FOR ENGINEERING
with Applications to Reliability

Chapter 1: The Basics

In January 1986 the space shuttle *Challenger* exploded in midair. The space shuttle had previously been considered so safe by NASA that plans were afoot to send plutonium powered modules into orbit. The previous year an accident at the Union Carbide plant in Bhopal, India, had killed thousands of people in what was at the time the worst industrial accident in history. A few months later, the Soviet reactor at Chernobyl was to run amok and burn out of control for days, while spewing radioactivity into the atmosphere. The Soviets had thought the probability of such a major accident to be extremely low.

As long as such accidents continue, and there is every indication that they will, there will be a lot of interest in trying to determine the reliability of things like space vehicles, nuclear and chemical reactors, and such ordinary devices as automobiles, garage door openers, or heating systems. Skepticism now greets claims that something or other is "less likely to happen to you than being hit by lightning," or has "only 1 chance in 10,000 of occurring in the next 50 years." Often such claims have been based more on hope than on science.

Estimates of the reliability of equipment or complex systems depend heavily on the field of mathematics known as *probability*. Probability can be abused, just as can most tools. The best mathematical model can't produce true answers if incorrect or naive assumptions are fed into it. Even at a fairly elementary level, however, probability opens the door to the investigation of complex systems and situations. If we want to answer such questions as "What were the chances of that happening?" or "How much do we expect to gain if we make that decision?", the answer will have to be expressed in the language of probability. The purpose of this book is to present the basics of that language and to show its application to a variety of meaningful examples, with an emphasis on the idea of reliability.

Interest in probability blossomed around the gambling tables of Europe hundreds of years ago, though much earlier references can be found in Hebrew and Chinese. Many games can be analyzed by looking at the possible outcomes of an experiment, such as rolling a pair of dice or dealing some cards from a deck. Frequently something about the situation suggests that the various possible outcomes should be considered equally likely. For example, the symmetric shape of a six-sided die suggests that the six outcomes are equally probable, and the purpose in shuffling a deck of cards before dealing is to try to approximate a situation in which one arrangement of the deck is just as likely as another.

The concept of equally likely outcomes leads to a natural concept of probability. For example, since there are 13 hearts in a deck of cards, there are 13 chances out of 52 that an arbitrary card drawn from a deck will be a heart. Considering the ratio of these two numbers gives the intuitively satisfying conclusion that the probability of drawing a heart when a card is drawn from a deck should be 13/52, or 1/4.

A more skeptical person might argue that the only way to test probabilities is to experiment. For example, given a coin of unknown characteristics, the only way to determine the probability of the coin coming up heads is to toss it many times and see what happens. A mathematician might look at the situation in this way: If H_n is the number of heads obtained in the first n tosses of a sequence of tosses, the probability of heads might be taken as

$$\lim_{n\to\infty} \frac{H_n}{n}$$

Of course there's no way actually to toss a real coin an infinite number of times to evaluate the limit, but the intuitive idea is that such a limit ought to exist and should define whatever it is we mean by the probability of the coin coming up heads.

A third way that probabilities are tossed about in everyday conversation involves subjective considerations. Someone might say, for example, "Notre Dame is a 2 to 1 favorite to beat Michigan." This statement has a clear meaning as far as probabilities go. The speaker is saying that Notre Dame's chances of winning are 2 chances out of 3, which means a probability of 2/3. Such a statement is the speaker's quantitative pronouncement of his or her or somebody's opinion on the matter. Another such illustration is a weather forecaster announcing a 30% chance of rain. Presumably such a statement would be based on existing weather data and would not be purely subjective. It may be, however, that some kind of subjective guesswork went into building the weather model from which the 30% figure is obtained.

A mathematically useful treatment of probability must lay a common groundwork so that everyone is speaking the same language. Much of this groundwork consists of the elementary language of *sets* and *set operations*.

1.1 Sets and Set Operations

Intuitively, a set is simply a collection of objects. This is one of the most primitive of mathematical concepts, and thus we cannot define sets in terms of yet more elementary concepts. The common practice is to denote sets by capital letters.

Equality of two sets means that the sets consist of exactly the same elements. For example, $A = \{1, 2, 3\}$ and $B = \{2, 3, 1\}$ are equal. (There is no order associated with the elements of a set. A set is simply an unordered collection of objects.) If every element of set A is also an element of set B, then A is a *subset* of B and we write $A \subset B$. Set membership is denoted by the symbol $\in$. For example, $1 \in A$ but $4 \notin A$ in this example.

Perhaps the most useful method of defining sets is by describing the rule for set membership. For example,

$$S = \{x : x \text{ is an integer and } x > 0\}$$

is simply a way of describing S as the set of positive integers.

The three basic set operations are union, intersection, and complementation. Union and intersection are operations that are performed on collections of sets, whereas complementation is performed on a single set.

The *union* of two sets A and B is denoted by $A \cup B$ and defined by

$$A \cup B = \{x : x \in A \text{ or } x \in B\}$$

It is important to understand that "or" allows the possibility of membership in both. Thus the condition for membership in $A \cup B$ is simply that the candidate be an element of at least one of the two sets A and B.

In the definition of the *intersection* of two sets, "or" becomes "and." Thus,

$$A \cap B = \{x : x \in A \text{ and } x \in B\}$$

Intersection thus represents the "overlap" of the two sets. If this intersection is empty, then the sets are said to be *disjoint* or *mutually exclusive*. Both the definition of union and intersection extend naturally to any collection of sets (even an infinite collection). The *union of a collection of sets* consists of all elements that belong to at least one set of the collection, whereas the *intersection of the collection* consists of the elements that all the sets have in common.

In any discussion about sets, the sets under consideration will all be subsets of some *universal set* lurking in the background. (What the universal set is should always be clear from context.) The *complement* of a set A is denoted by A^c and is defined as the set of all elements in the universal set that do *not* belong to A. For example, in the context of a discussion of the real number system, the complement of $A = \{x : x > 1\}$ is

$$A^c = \{x : x \leq 1\}$$

The *difference* of two sets, $A - B$, is defined as $A \cap B^c$.

Some of the important elementary laws governing the set operations are as follows:

Associative law for union	$A \cup (B \cup C) = (A \cup B) \cup C$
Associative law for intersection	$A \cap (B \cap C) = (A \cap B) \cap C$
Commutative law for union	$A \cup B = B \cup A$
Commutative law for intersection	$A \cap B = B \cap A$
Distributive laws	$A \cap (B \cup C) = (A \cap B) \cup (A \cap C)$
	$A \cup (B \cap C) = (A \cup B) \cap (A \cup C)$
De Morgan's laws	$(A \cup B)^c = A^c \cap B^c$
	$(A \cap B)^c = A^c \cup B^c$

A simple way to verify some elementary set identities is to use *Venn diagrams*. The idea is to visualize a set as being represented by a region in the plane. Figure 1.1 illustrates the use of this concept with regard to the first distributive law above. The shaded area in the figure is the region corresponding to the sets on either side of the equation in the first distributive law. One way to check equality of sets is to construct a Venn diagram for each set and to observe that the Venn diagrams coincide.

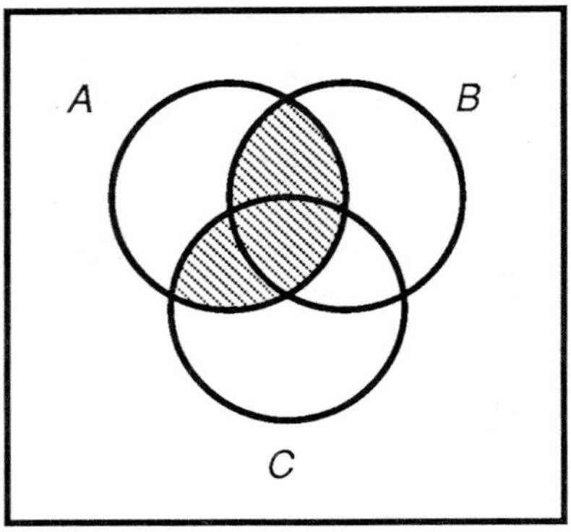

Figure 1.1 Venn diagram representing the set $A \cap (B \cup C) = (A \cap B) \cup (A \cap C)$.

Example 1.1. A group of 9 men and 7 women are administered a test for high blood pressure. Among the men, 4 are found to have high blood pressure, whereas 2 of the women have high blood pressure. Use a Venn diagram to illustrate this data.

Solution:

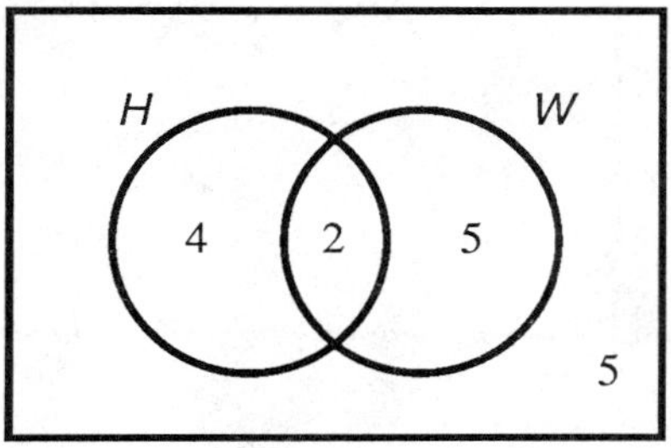

The circle labeled H represents the 6 people having high blood pressure, and the circle labeled W represents the 7 women. The numbers placed in the various regions indicate how many people there are in the category corresponding to the region. For example, there are 4 people who have high blood pressure and who are not women. Such people are in set H but not in set W; that is, they belong to the set $H \cap W^c$. The number 5 in the lower right corner indicates the number of men without high blood pressure.

The decision to use circles to represent "high blood pressure" and "women" was quite arbitrary. We could just as well use circles for "low blood pressure" and "men." (See Problem 1.25.)

Example 1.2. If A, B, and C are sets, draw a Venn diagram and shade the region corresponding to the set $(A \cup B^c) \cap C$.

Solution: The best way to arrive at the following figure is in a step-by-step manner. First, decide what region corresponds to the set $A \cup B^c$. This will include all the space *inside* the A circle and all the space *outside* the B circle. Then, the final step is to observe what part of this region lies *inside* of C (since the final operation is intersection). If you like you can think of spray painting everything inside the A circle, then spray painting everything outside the B circle, and then looking to see what part of the inside of circle C is painted. The region that represents the set $(A \cup B^c) \cap C$ is shown in the figure.

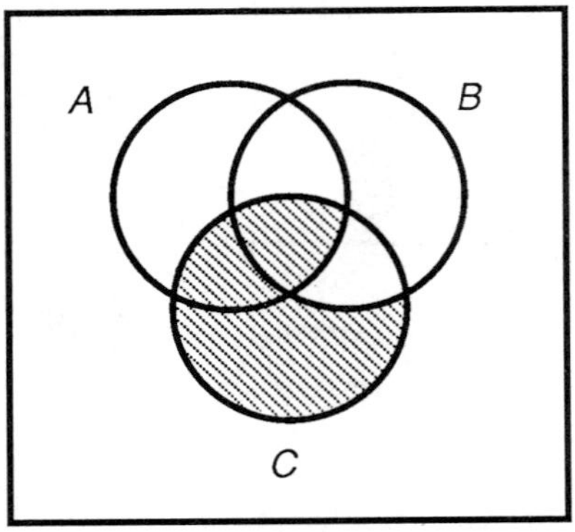

Example 1.3. If A, B, and C are the following sets of characters, then determine the set $(A \cup B^c) \cap C$. Here we will consider the universal set to be *all* 26 letters of the alphabet.

$$A = \{a, h, t\}$$
$$B = \{d, g, b, t\}$$
$$C = \{d, a, g\}$$

Solution: While it is possible to represent this information in a Venn diagram, it certainly isn't necessary. Simply observe that B^c consists of all letters except d, g, b, and t. Therefore, $A \cup B^c$ consists of all letters except d, g, and b. So the only letter that $A \cup B^c$ has in common with C is a. Conclusion: $(A \cup B^c) \cap C = \{a\}$. Notice that Problem 1.27 asks you to observe this in the context of a Venn diagram.

1.2 The Sample Space

The *sample space* is roughly "the set of all possible observations or outcomes" of whatever is under discussion. This idea is best illustrated through examples.

Example 1.4. If a coin is tossed, the sample space could be taken to be the set $S = \{\text{H}, \text{T}\}$. If an ordinary six-sided die is rolled, the sample space could be taken to be the set $S = \{1, 2, 3, 4, 5, 6\}$.

Example 1.5. A card is drawn from a standard deck of 52. Here one could take the sample space S to be

$$S = \{2\clubsuit, 2\diamondsuit, 2\heartsuit, 2\spadesuit, 3\clubsuit, 3\diamondsuit, \cdots, K\spadesuit, A\clubsuit, A\diamondsuit, A\heartsuit, A\spadesuit\}$$

A simpler convention would be to agree to think of the cards as being identified with the numbers 1,2, $\cdots$, 52 and to simply think of S as consisting of these 52 numbers. It is important to realize that the particular bookkeeping scheme used is not very important compared to the conceptual understanding of what kind of set is an appropriate model. Whatever notation is used here, the sample space is a set of 52 elements.

Example 1.6. Suppose a simple electric circuit has two components, say A and B. Either component can be "good" or "bad" in the sense that the component may or may not be in working order. If we are interested in all possible states of the circuit, the sample space used could be S = {GG, GB, BG, BB}, where the convention might be that "GB," for example, means that component A is good and component B is bad.

Example 1.7. A pair of dice is rolled. Let's refer to them as "red" and "green." An appropriate choice for the sample space for this experiment would be the set $S = \{(1, 1), (1, 2), (1, 3), \cdots, (6, 5), (6, 6)\}$, where, for instance, we might agree that (3, 5) represents the outcome of 3 on red and 5 on green. The set S contains 36 elements since either die can come up 6 different ways and 6×6 is 36. (There's a basic underlying principle here that some elementary texts call the *multiplication principle.* When one task can be performed in m different ways and another task can be performed in n different ways, then the number of different ways of performing the two operations together is mn.)

In playing many games, one is not interested in the individual numbers that appear on the dice, but rather in the sum. In this case it might be tempting to take the sample space to be $S = \{2, 3, 4, \cdots, 11, 12\}$. This is not necessarily wrong, but it does sacrifice information. For example, if this sample space is used, one can no longer answer questions such as "Did the red die show an even number?" The result on the red die is not even being recorded. Another reason for caution is that the outcomes of this set of possible sums are not equally likely. For example, a sum of 2 occurs only if both dice show the number 1, whereas a sum of 7 occurs in six different ways. It is often necessary to consider sample spaces in which individual outcomes don't all have the same probability. One needs, however, to be aware when this is the situation. A frequent naive mistake is to assume that outcomes are equally likely when they are not. It is common to refer to a sample space in which all the elements are considered equally probable as a *uniform sample space*.

Example 1.8. When binary data is transmitted, the output can be thought of as a string of 0's and 1's. (In electrical transmission, a voltage above a certain level could be defined as 1 and below a certain level as 0.) If 4 bits are transmitted, what would be an appropriate sample space to represent the possibilities?

Solution: The logical choice would be to take the sample space to be all ordered quadruples of binary digits. In other words, S = {0000, 0001, 0010, 0011, 0100, 0101, 0110, 0111, 1000, 1001, 1010, 1011, 1100, 1101, 1110, 1111}. These 16 outcomes are simply the numbers from 0 to 15 written in binary form. There are 16 elements of the sample space because $2^4 = 16$. Eight bits (commonly referred to as a byte) can represent $2^8 = 256$ different possibilities. This is equivalent to saying that if 8 bits are transmitted, then the sample space for all possible outcomes has 256 elements.

1.3 Basic Properties of Probabilities

The word *event* is commonly used in everyday language, and often we speak of a particular event "occurring." In probability discussions, you should think of an *event* as a subset of the sample space. For example, if we say when a die is rolled that "an even number occurs," we are saying that the observed outcome lies in the set $E = \{2, 4, 6\}$. We are describing E verbally by saying E is the event "that an even number occurs," but this is simply an alternate way of saying that E is the given set of outcomes. The reason for talking about events occurring is that this language is so common and useful in everyday speech. In talking about probability, to say that an event occurs is simply to say that the observed outcome is one of the elements of the event. This mathematical language is consistent with common speech. You should keep in mind, though, that in the language of probability there is a precise mathematical meaning to such expressions.

There are three basic axioms that characterize what is meant by a *probability measure*. The notation $P(A)$ represents the probability that the event A occurs, and the assumption is that probabilities always behave according to the these rules.

The Basic Axioms for a Probability Measure

1. For each event A, $P(A) \geq 0$.
2. $P(S) = 1$, where S denotes the sample space.
3. If A and B are disjoint, then $P(A \cup B) = P(A) + P(B)$.

The intuitive basis for all three axioms should be obvious. When we say that something has probability 1/10, we mean that there is one chance in 10 that it will occur. A negative probability has no conceivable meaning. Similarly, a probability of 1 represents absolute certainty, and probabilities greater than 1 would be meaningless. The last axiom is very important in that it allows the probability of events (in the case of a finite sample space) to be computed in terms of the probabilities of the individual elements that make up the events. This will be demonstrated shortly. (See Equation 1.1.)

Additional Properties of Probabilities

1. $P(\emptyset) = 0$, where $\emptyset$ denotes the empty set.
2. $P(A) \leq P(B)$ if $A \subset B$, and $P(A) \leq 1$ always.
3. $P(A - B) = P(A) - P(A \cap B)$.
4. $P(A \cup B) = P(A) + P(B) - P(A \cap B)$ always.
5. If $A_1, A_2, \cdots, A_n$ are disjoint events (no two events having any elements in common), then

$$P(A_1 \cup A_2 \cup \cdots \cup A_n) = P(A_1) + \cdots + P(A_n)$$

Property 1 is an immediate consequence of Axiom 3. (Simply take A and B both to be the empty set in Axiom 3, and you have $P(\emptyset) = 2\,P(\emptyset)$, which implies $P(\emptyset) = 0$.)

If $A \subset B$ then $B = A \cup (B - A)$, and this is a disjoint union. Thus, from Axiom 3 we know that $P(B) = P(A) + P(B - A)$. The right side here is greater than or equal to $P(A)$ because $P(B - A) \geq 0$ (Axiom 1), and this proves Property 2.

Property 3 is a result of the fact that $A = (A \cap B) \cup (A - B)$, and the union here is disjoint.

Property 4 may be obtained by first noting that

$$P(A \cup B) = P(A) + P(B - A)$$

This is a special case of Axiom 3. From Property 3, however, we know that

$$P(B - A) = P(B) - P(A \cap B)$$

The truth of Property 5 is a consequence of mathematical induction. In fact, if

$$P(A_1 \cup A_2 \cup \cdots \cup A_k) = P(A_1) + \cdots + P(A_k)$$

then it follows that

$$P(A_1 \cup A_2 \cup \cdots \cup A_{k+1}) = P(A_1) + \cdots + P(A_{k+1})$$

simply by letting $A_1 \cup A_2 \cup \cdots \cup A_k$ and A_{k+1} play the roles of A and B in Axiom 3.

For finite sample spaces it is always possible to compute the probability of an event by focusing on the individual elements that make up the event. For if the sample space is $S = \{s_1, s_2, \cdots, s_n\}$, then, according to Property 5, the probability of any event $A \subset S$ may be computed via

$$P(A) = \sum_{s_k \in A} P(\{s_k\}) \tag{1.1}$$

In other words, the probability of any (finite) event may be computed by simply adding up the probabilities of the individual elements of the event.

Later in this book we will see that probabilities of events often must be approached from a different perspective when the sample space is infinite. The idea of computing probabilities of all events via a sum, as in Equation 1.1, must be abandoned. In Chapter 5 we will see that in "continuous" models, integration takes the place of summation.

Continuous models arise, for instance, when measurements are being made on some kind of continuous scale and one wishes to think of an interval of possible values. For example, think of the experiment of "selecting a random number between 0 and 1." Clearly it would be intuitively satisfying to think that the probability is 1/2 that the number should come from the subinterval [0, 1/2], or 1/5 that the number chosen should be in the interval [3/5, 4/5]. In fact, we would like to know that the probability of the number falling in any particular subinterval is simply the length of the subinterval. Simple as this situation sounds, the actual demonstration that there is a probability measure on the interval [0, 1] that has these pleasant properties was a milestone in mathematics around the turn of the century and made possible the field of study known as *real variables*. An interesting consequence of the Axioms 1 through 3 is that such a probability measure must necessarily assign probability zero to every individual number. To understand why, simply think of a number x sitting in the interval $B_\delta = [x - \delta, x + \delta]$. The length of this interval is 2δ. But in light of Property 3, $P(\{x\}) < P(B_\delta)$ for every $\delta > 0$, and so $P(\{x\})$ must be zero. (To say that an event has probability zero is not to say that the event is impossible. For instance, if a random number is selected between zero and one, the probability that the number would be 1/2 is zero as just indicated. Yet it is not impossible for the number selected to be 1/2.)

For the sake of honesty, it is advisable to admit at this point that there are still other pathological things that occur with infinite sample spaces. With finite sample spaces, an *event* is simply a subset of the sample space. With infinite sample spaces, it is not always possible to consider *every* subset of the sample space to be an event. The problem is that for infinite sample spaces it simply isn't usually possible to have our probability measures defined for *all* subsets of the sample space and still to satisfy the desired axioms. You needn't lose sleep over this, however, because in all the frequently encountered sample spaces, the subsets that must be avoided are ones that you would never encounter anyway because they cannot be expressed in terms of elementary sets. In other words, you can ignore this conceptual difficulty and never run into computational problems because of it.

1.4 Conditional Probability

New information changes one's notion of probabilities. For example, a card drawn from a deck has probability 1/4 of being a spade since 1/4 of the cards in the deck are spades. However, if it is known that the card is a black card, then the *conditional probability* that the card is a spade becomes 1/2 since half of the *black* cards are spades. Effectively, what has happened is that the additional information (that the card drawn is a black card) has eliminated some of the original possibilities. This forces one to reconstruct the sample space to form a new "reduced" sample space consisting of only 26 of the original 52 possible outcomes. The definition of conditional probability given below represents a slightly different view. The examples that follow indicate that the definition is consistent with the intuitive view of a "reduced sample space."

Definition 1.1: If A and B are events in a sample space and $P(A) \neq 0$, then the conditional probability of B, given A, is denoted by $P(B \mid A)$ and is defined by

$$P(B \mid A) = \frac{P(B \cap A)}{P(A)}$$

Figure 1.2 shows a graphical interpretation of the idea of conditional probability. Both probability and area can be thought of as measures of size. In fact, $P(B \mid A)$ is a measure of the size of $P(B \cap A)$ in comparison to $P(A)$. In Figure 1.2, the analogous idea would be to compare the area of $B \cap A$ to the area of A. In both the probability setting and the geometric figure, we are in some sense measuring what fraction of A happens to also lie in the set B.

When considering $P(A \mid B)$, the size of $P(B \cap A)$ is compared to $P(B)$ rather than to $P(A)$. Quite possibly $P(B \mid A)$ and $P(A \mid B)$ may be quite different in magnitude. In the figure the region $B \cap A$ constitutes a much larger fraction of A than it does of B.

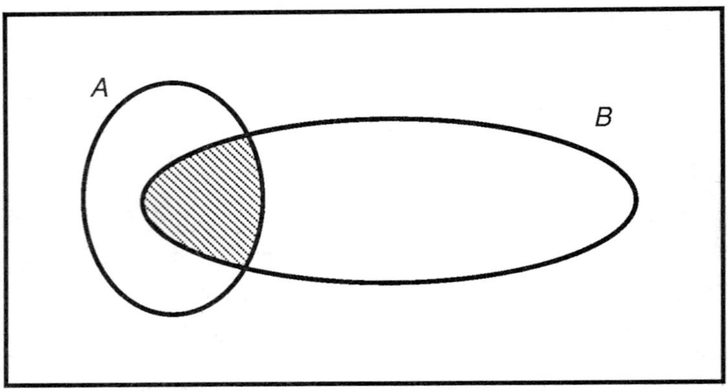

Figure 1.2 How probable is $A \cap B$ in comparison to A? This is the idea behind the conditional probability $P(B \mid A)$.

Example 1.9. A card is drawn from a 52 card deck. Let A be the event that the card is black and B the event that the card is a spade. Thus $A \cap B = B$, so $P(A \cap B) = 13/52$, and $P(A) = 26/52$. This means that

$$P(B \mid A) = \frac{13/52}{26/52} = \frac{1}{2}$$

Example 1.10. A pair of dice is rolled (one red and one green). The sample space is as in Example 1.7. Let A be the event that the sum on the two dice is 9, and let B be the event that the red die shows the number 5. Then $A \cap B$ contains the single outcome (5, 4) and has probability 1/36, and $P(A) = 4/36$ because $A = \{(3, 6), (4, 5), (5, 4), (6, 3)\}$. Thus, $P(B \mid A) = 1/36 \div 4/36 = 1/4$. This is intuitively correct, for knowing that the sum is 9 guarantees that the red die must show one of the four numbers 3, 4, 5, or 6.

Example 1.11. In the sample space of Example 1.5, let B denote the spades and now let C denote the aces. Then $P(B \cap C) = 1/52$ since there is only one card that is the ace of spades, and so $P(B \mid C) = 1/52 \div 4/52 = 1/4$. Notice that in this case $P(B \mid C) = P(B)$. It can be easily verified that in this example it is also true that $P(C \mid B) = P(C)$. These two equations indicate that the knowledge that one of the events occurs does not affect the probability of the other occurring. Two events that have this relation to each other are called *independent events* (or *statistically independent events*). The concept of independence is very important in mathematical models that involve probability. In many situations it will be apparent for one reason or another that two events can have no possible effect on one another, and in the mathematical model this leads to an assumption that the events are independent.

Example 1.12. Two bits of binary data are transferred. Let's consider the sample space to be S = {00, 01, 10, 11}. So 01 represents the case in which a 0 is transmitted first and a 1 is transmitted as the second bit. We will consider the transmission to be random in the sense that each of these arrangements has probability 1/4.

Now let's consider the following two events:

$$A = \{10, 11\}$$
$$B = \{01, 11\}$$

Event A can easily be described as the event "the first digit transmitted is the digit 1," and B as the event "the second digit is the digit 1."

Notice that $P(B) = 1/2$, and furthermore that $P(B \mid A) = 1/2$. (Why?) The fact that $P(B) = 1/2$ says that the probability that the second digit is a 1 is equal to

1/2. That $P(B \mid A) = 1/2$ says that if we know that the first digit is a 1, then the conditional probability (based on this information) that the second digit is a 1 is *still* 1/2. This is another illustration of independence, the concept introduced in Definition 1.2 below. Problem 1.28 asks you to show that the three conditions are in fact equivalent (that is, if one of them is true, then all of them have to be true).

Definition 1.2: Given events A and B with nonzero probability, it is easy to check that the following three conditions are equivalent:

1. $P(A \mid B) = P(A)$
2. $P(B \mid A) = P(B)$
3. $P(A \cap B) = P(A)\, P(B)$

When these conditions are true, the events A and B are said to be *independent.*

Conditional probabilities have many uses. Strangely enough, one of the easiest and most useful is in the computation of unconditional probabilities. In fact, the definition of $P(B \mid A)$ can be rewritten as

$$P(A \cap B) = P(A)\, P(B \mid A) \tag{1.2}$$

This idea can be extended to compute the occurrence of more than two events. For example,

$$P(A \cap B \cap C) = P(A)\, P(B \mid A)\, P(C \mid A \cap B) \tag{1.3}$$

Now for some examples that illustrate the use of these simple but useful little formulas.

Example 1.13. A box contains 10 transistors, of which 7 are good and 3 are defective. If 2 transistors are randomly taken from the box, what is the probability that both are good?

Solution: Think of the transistors as being drawn one at a time. (This just means that we're going to label them "first" and "second." Whether they are actually drawn one at a time or simultaneously doesn't matter in the least. You should convince yourself that this little mental trick is legitimate.) Let A be the event that the first is good, and B the event that the second is good. Then the probability that both are good is

$$P(A \cap B) = P(A)\,P(B \mid A) = \frac{7}{10} \times \frac{6}{9} = \frac{7}{15}$$

The reason that $P(B \mid A) = 6/9$ is that knowing that the first one is good means that there are 6 good left among the 9 possible ones that might be chosen second.

In a similar fashion, the probability that all would be good if 3 were selected could be computed as

$$\frac{7}{10} \times \frac{6}{9} \times \frac{5}{8}$$

The last factor, 5/8, is the conditional probability that the third would be good given that the first two are both good.

Example 1.14. Consider the experiment of rolling a red and a green die. Let A be the event that the red die shows a 5, and B be the event that the green die shows a 3. Most people's intuition says that the two dice should not exercise any control over each other; that is, the outcome on one die should be "independent" of the outcome on the other. If the sample space of Example 1.7 is used, then

$$A \cap B = \{(5, 3)\}, \text{ so } P(A \cap B) = \frac{1}{36}$$

Also, $B = \{(1, 3), (2, 3), (3, 3), (4, 3), (5, 3), (6, 3)\}$, and so $P(B) = 6/36$.

Thus,

$$P(A \mid B) = \frac{1/36}{6/36} = \frac{1}{6} = P(A)$$

This shows that A and B are independent according to the Definition 1.2, just as our intuition would indicate. (A somewhat subtle but important point should be made here. One cannot prove mathematically whether two real-world dice behave independently or not. All we are illustrating is that if the sample space introduced in Example 1.7 is used to model a pair of dice, with the assumption that the outcomes are equally likely, then the theoretical dice of the model behave independently.)

Example 1.15. A manufacturing company has two plants, 1 and 2. Plant 1 produces 40% of the company's output, and plant 2 produces the other 60%. Of the devices produced at plant 1, 95% are good and 5% are defective. The output of plant 2 is 90% good and 10% defective. If a device is randomly selected from the output of this company, what is the probability that the device will be good?

Solution: Let B denote the event that the randomly selected item is good, and let A_1 and A_2 be the events that it comes from plants 1 and 2 respectively. Then $P(B) = P(B \cap A_1) + P(B \cap A_2)$ because A_1 and A_2 are disjoint events and every element of B must be in either A_1 or A_2. But then

$$\begin{aligned} P(B) &= P(B \cap A_1) + P(B \cap A_2) \\ &= P(A_1)\,P(B \mid A_1) + P(A_2)\,P(B \mid A_2) \\ &= (.4)(.95) + (.6)(.9) = .92 \end{aligned}$$

(We are using Equation 1.2 here. An alternative is to use a *tree diagram* for this computation. Tree diagrams are explained in Section 1.5.)

Example 1.16. Suppose it is known that 3% of cigarette smokers develop lung cancer, whereas only .5% of nonsmokers develop lung cancer. Furthermore, let's suppose that 30% of adults smoke. If it is known that a randomly chosen adult has developed lung cancer, what is the probability that the person is a smoker?

Solution: Let S = smokers and L = people who develop lung cancer. The question is what is $P(S \mid L)$. By definition $P(S \mid L) = P(S \cap L)/P(L)$. However,

$$P(S \cap L) = P(S)\,P(L \mid S) = (.3)(.03) = .009$$

Moreover,

$$\begin{aligned} P(L) &= P(S \cap L) + P(S^c \cap L) \quad \text{since } S \cap L \text{ and } S^c \cap L \text{ are disjoint} \\ &= P(S)\,P(L \mid S) + P(S^c)\,P(L \mid S^c) \\ &= (.3)(.03) + (.7)(.005) = .009 + .0035 = .0125. \end{aligned}$$

Thus $P(S \mid L) = .009/.0125 = .72$.

The technique used in this computation may be incorporated in a tree diagram or Bayes' theorem, both introduced in the next section. (See Problem 1.21.)

1.5 Tree Diagrams and Bayes' Theorem

The techniques used in Examples 1.15 and 1.16 can be formalized in two different ways. One formulation is commonly referred to as Bayes' theorem and is included in Proposition 1.1. Another approach utilizes a simple technique called *tree diagrams* which will be introduced shortly.

Bayes' theorem comes into play whenever the elements of the sample space under consideration are divided up into mutually exclusive categories. Specifically, suppose that S is the sample space and that $A_1, A_2, \cdots, A_n$ are events in S having the property that $A_1, A_2, \cdots, A_n$ are disjoint and that $A_1 \cup \cdots \cup A_n = S$. (This means that each element of the sample space S belongs to exactly one of the events $A_1, A_2, \cdots, A_n$, and for that reason such a collection of sets is often referred to as a *partition* of the sample space.) Then for any event $B \subset S$, the additivity property of probabilities implies that

$$P(B) = P(B \cap A_1) + \cdots + P(B \cap A_n)$$

If each of the terms on the right is rewritten using the definition of conditional probability, the first equation in Proposition 1.1 is obtained (the multiplicative law). Bayes' theorem itself is an immediate consequence of the multiplicative law. The definition of conditional probability says that

$$P(A_k \mid B) = \frac{P(A_k \cap B)}{P(B)}$$

If the definition of conditional probability is used now to rewrite the numerator of this fraction, and if the multiplicative law is used to replace $P(B)$ in the denominator by the sum on the right side of Equation 1 in Proposition 1.1, then Equation 2 (Bayes' theorem) is obtained.

Proposition 1.1: Suppose that B and $A_1, A_2, \cdots, A_n$ are events in a sample space S, that $A_1, A_2, \cdots, A_n$ are disjoint, and that $A_1 \cup \cdots \cup A_n = S$.

1. (multiplicative law) Then for any event B in S,

$$P(B) = \sum_{i=1}^{n} P(A_i)\, P(B \mid A_i)$$

2. (Bayes' theorem) For any one of the sets A_k,

$$P(A_k \mid B) = \frac{P(A_k)\, P(B \mid A_k)}{\sum_{i=1}^{n} P(A_i)\, P(B \mid A_i)}$$

The solution shown in Example 1.16 is exactly the one that would be produced by using Bayes' theorem. To make the identification, simply let the event L in Example 1.16 correspond to B in Bayes' theorem, and let S and S^c correspond to A_1 and A_2 (with $n = 2$ and $k = 1$) in Bayes' theorem.

Many simple problems that can be done with Proposition 1.1 can also be done with tree diagrams. The concept of a tree diagram is easily understood by looking at a few examples.

Example 1.17. To check for mercury contamination, a group of 25 dolphins is being monitored. Of the 25 dolphins, 6 have been determined to contain a high concentration of mercury contamination in their fatty tissues, whereas the other 19 do not. If 2 dolphins are randomly chosen from this group of 25, what is the probability that both will have high mercury concentrations?

Solution: The purpose of this example is to introduce the concept of a tree diagram. The tree that is constructed will show show more information than is required to answer the question being posed.

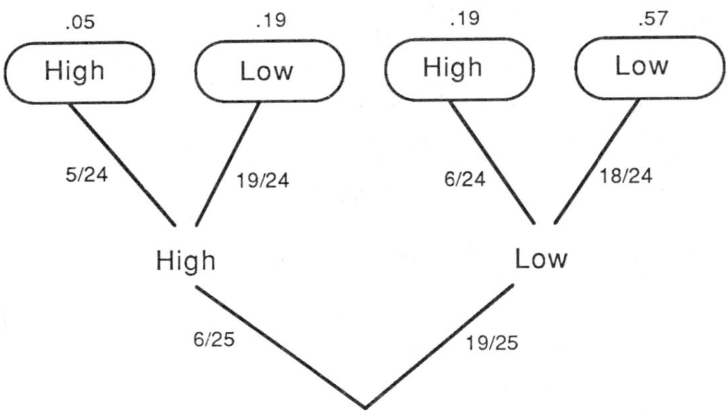

At the lower level in the tree, the nodes labeled "High" and "Low" represent the two possibilities (high or low mercury concentration) for the *first* dolphin. (Again, it doesn't really matter whether they are chosen one at a time or simultaneously. We can simply think of one of them as being labeled "first" and the other as being labeled "second.") The probabilities on the branches are 6/25 and 19/25 because of what is known about the makeup of the group of 25. At the second level in the tree, the possible conditions of the second dolphin are recorded. The numbers on the second level of branches are conditional probabilities. For example, the leftmost number 5/24 at the second level is the conditional probability that the second dolphin has a high concentration of mercury given that the first one does. The reason this conditional probability is 5/24 is that once we know that the first one has a high concentration, we know that 5 of the remaining 24 also have high concentrations. The four ovals at the top of the figure represent the four possible outcomes of the experiment. From left to right they may be described as: both high, first high and second low, first low and second high, both low. Thus the tree diagram is consistent with thinking of the sample space as something like $S = \{HH, HL, LH, LL\}$. Finally, the numbers at the top of the figure are the probabilities of the four outcomes corresponding to the ovals. These are computed by multiplying the numbers on the branches connecting the root of the tree to the ovals. For example, the probability of the outcome HH is computed as $6/25 \times 5/24 = .05$. This use of multiplication is an application of Equation 1.2.

Example 1.18. Two bad resistors have been inadvertently installed in a device having five resistors altogether. It is not known which two are bad. Resistors are tested one at a time until the two bad ones or the three good ones have been found. Draw a tree diagram to represent this process, and find the probability that exactly two good resistors are tested.

Solution: In the following figure "G" represents selection of a good resistor and "D" represents selection of a defective one. The first level of the tree represents the first resistor selected, and the probabilities shown there are 3/5 and 2/5, corresponding to the makeup of the set of five resistors. Above the lowest level the probabilities are conditional probabilities. The three shaded circles represent the outcomes in which exactly two good resistors are tested. The desired probability is then just the sum of the probabilities of these three outcomes. As before the probabilities of outcomes may be computed by multiplying the numbers on the branches between the root and the top. In this example, all outcomes turn out to have probability 1/10, and so the probability that exactly three good resistors are tested is 3/10.

Since the tree diagram shows all the relevant information about the experiment, conditional probabilities also can be computed from the tree. For example, suppose A is the event that both defective resistors are tested, and B is the event that exactly three resistors are tested altogether. In the diagram there are three outcomes in which exactly three resistors are tested (the circles at the intermediate level). Since each has probability 1/10, $P(B) = 3/10$. $A \cap B$ consists of two of these outcomes; so $P(A \cap B) = 2/10$. Therefore,

$$P(A \mid B) = \frac{2/10}{3/10} = \frac{2}{3}$$

that is, if it is known that exactly three resistors are tested, then the probability that both defectives are tested is 2/3.

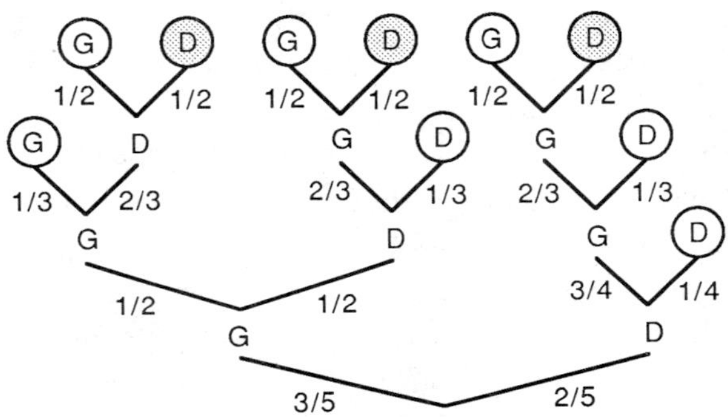

It is very important to understand clearly the difference between the conditional probability $P(A \mid B)$ and $P(A \cap B)$. Beginning students sometimes confuse these two probabilities. Problem 1.11 at the end of the chapter is a good check on this concept.

1.6 More on Independence

Sometimes it is necessary to consider the independence of more than two events. For three events, it would be tempting to want to consider events A, B, and C to be independent if $P(A \cap B \cap C) = P(A)\, P(B)\, P(C)$. It is easy to see that this view is inadequate, however, by considering the special case where C is empty. For then both sides of the equation are zero, and so the equation is necessarily valid even though A and B may be highly dependent. The following definition is what is needed.

Definition 1.3: A collection of events is *independent* provided that for every finite subcollection $A_1, \cdots, A_n$ it is true that

$$P(A_1 \cap \cdots \cap A_n) = P(A_1) \cdots P(A_n)$$

For example, three events A, B, and C are independent provided all of the following are true:

1. $P(A \cap B \cap C) = P(A)\, P(B)\, P(C)$
2. $P(A \cap B) = P(A)\, P(B)$
3. $P(A \cap C) = P(A)\, P(C)$
4. $P(B \cap C) = P(B)\, P(C)$

Example 1.19. A coin is tossed twice. A natural sample space is given by $S = \{\text{HH, HT, TH, TT}\}$, with each outcome having probability 1/4. Consider the events:

$$A = \{\text{HH, HT}\} = \text{event that heads occurs on first toss}$$
$$B = \{\text{HH, TH}\} = \text{event that heads occurs on second toss}$$
$$C = \{\text{HH, TT}\} = \text{event that same result occurs on the two tosses}$$

It is routine to verify that equations 2 through 4 above are valid when A, B, and C are these three events. This simply says that any two of these three events are independent. However,

$$P(A \cap B \cap C) = P(A \cap B) = \frac{1}{4}$$

whereas

$$P(A)\, P(B)\, P(C) = \frac{1}{2} \times \frac{1}{2} \times \frac{1}{2} = \frac{1}{8}$$

So the collection of three events does *not* form an independent collection. One way to understand this is to observe, for example, that if we know that A and B occur, then we know with absolute certainty that C occurs. Intuitively, independence of the collection of events would indicate that no knowledge involving only A and B could influence the probability of C occurring. (A more mathematically precise statement is the following: If A, B, and C are independent, then C will be independent of any event that can be formed from A and B using set union, intersection and complementation. This can be proved using Definition 1.3.) In particular, *in a collection of independent events, any of the events may be replaced by their complements and the resulting collection of events will still be independent.* This will be extremely important to us in Chapter 2. Most of the applications will feature the concept of independent events in one form or another. In Chapter 2, we will constantly be performing calculations such as

$$P(A \cap B \cap C^c) = P(A)\,P(B)\,[1 - P(C)]$$

if A, B, and C are independent.

There is a common situation called an *independent trials process* that is closely connected with the concept of independent events. An independent trials process is a sequence of experiments in which the result of each trial is one of only two outcomes commonly called *success* and *failure*. Furthermore, it is required that the probability of the two outcomes remain the same on every trial. The easiest illustration is repeated coin tossing, where we might agree to call heads "success." (The choice of which outcome to call success is purely arbitrary.) We will look at an easy formula for the probability of a certain number of successes in a given number of trials. First, however, it is useful to present an elementary counting technique.

Proposition 1.2 (Combinations formula): If n is a positive integer and if $0 \le k \le n$, then the number of different possible sets of k objects that can be chosen from a set of n objects is given by

$$C(n, k) = \frac{n!}{k!(n-k)!}$$

The notation $n!$ is for n *factorial,* that is, $n! = n(n-1)(n-2) \cdots 1$. For example, $4! = 4 \times 3 \times 2 \times 1 = 24$. By convention $0! = 1$. The proof of the above counting principle is not difficult, but we will omit it because counting techniques are not central to our purpose. Example 1.20 presents a simple situation to which this formula can be applied, and Example 1.21 shows a related situation in which it is not directly applicable.

Example 1.20. If you have offered to give a friend his choice of any 3 books from 10 that you own, the number of different ways he can make his selection is

$$C(10, 3) = \frac{10!}{3!\,7!} = 120$$

Notice that the selection of 3 books to be given away also completely determines which 7 will be left behind. In other words, $C(10, 3) = C(10, 7)$. More generally, if $0 \le r \le n$, then $C(n, r) = C(n, n-r)$.

Example 1.21. In a club with 10 members, how many ways are there to fill a slate of officers consisting of a president, secretary, and treasurer? The answer is $10 \times 9 \times 8 = 720$. The difference in this example and the last is that here we are not just selecting 3 people from 10, but specifically 3 people for 3 separate positions. Utilizing the multiplication principle, you can think of this as 10 choices for president, then (after that choice has been made) 9 possibilities for secretary, and then 8 for treasurer. The assumption is that no one occupies two offices.

It is standard practice to refer to problems such as Example 1.20 as *combinations* problems. These are problems in which a set of objects is being selected but no ordering of the objects is being considered. Problems such as Example 1.21 are called *permutations* problems. In such problems the order of selection does come into play. In Example 1.21, we are essentially assigning an order to the three officers by matching them up to the three specific offices. Generally, the term *permutation* is used to refer to an ordered arrangement of some collection of objects. Permutations and combinations problems are part of a field of mathematics known as *combinatorics*.

Proposition 1.3 (Independent Trials Formula): Suppose n trials are conducted of an experiment in which the probability of success on each trial is p and the probability of failure is $q = 1 - p$. Then for any integer k that satisfies $0 \le k \le n$, the probability of exactly k successes in the n trials is

$$C(n, k)\, p^k\, q^{n-k}$$

For example, suppose a regular unbiased six-sided die is rolled 4 times. What is the probability of obtaining two 6's? Considering a 6 to be a success, then

$p = 1/6$ and $q = 5/6$. Here $n = 4$ and $k = 2$ in Proposition 1.3. Therefore, the probability of exactly two 6's is

$$C(4, 2)\left(\frac{1}{6}\right)^2\left(\frac{5}{6}\right)^2 = \frac{25}{216}$$

You should verify the validity of the formula in this simple case by looking at the same experiment in terms of a tree diagram, with one level of branching for each roll of the die (with the outcomes for each roll shown as "6" or "not 6"). Doing this, in fact, leads to understanding why the formula is correct. For in a tree diagram to represent n independent trials, there would be exactly $C(n, k)$ outcomes across the top corresponding to k successes in the n trials, and each of these outcomes would have probability $p^k q^{n-k}$.

In the tree diagram for an independent trials process, the probability for success is the same at all locations in the tree. It is important to keep in mind that much more general kinds of experiments can be viewed in terms of a tree diagram than can be viewed as an independent trials process. The relation between an independent trials process and the concept of independent events is this: In an independent trials process, if A is an event that may be described in terms of the outcome of one trial and B is an event that may be described in terms of the outcome of another trial, then A and B will be independent events. This statement may be generalized to more than two events. For example, if a coin is tossed 10 times, the experiment may be viewed as 10 trials in an independent trials process. If A_1 is the event that "heads occurs on the first toss," A_2 that "heads occurs on the second toss," and so forth, then $A_1, \cdots, A_{10}$ are independent events.

1.7 Infinite Sample Spaces

When dealing with infinite sample spaces, it is often necessary to consider unions of infinite collections of events. For this reason, Axiom 3 is inadequate for infinite sample spaces and must be strengthened to Axiom 3.1 following. The reason is that mathematical induction cannot extend the property of finite additivity to the infinite series needed in the case of an infinite sample space. The right side of the equation in Axiom 3.1 is the sum of an infinite series. (The series will necessarily be convergent if the events are disjoint as in Axiom 3.1.)

Axiom 3.1: If $A_1, A_2, \cdots$ is a sequence of disjoint events in a sample space, then

$$P(A_1 \cup A_2 \cup \cdots) = P(A_1) + P(A_2) + \cdots$$

Axiom 3.1 simply says that even with a countably infinite collection of disjoint events, the probability of the union is still the "sum of the probabilities" provided that the latter is interpreted as the sum of an infinite series.

The concept of a *geometric series* will be useful in the next example and in other applications later.

Proposition 1.4 (Geometric Series): If $|r| < 1$ then

$$1 + r + r^2 + r^3 + \cdots = \frac{1}{1-r}$$

The derivation of the sum of a geometric series is easy, because

$$(1-r)\,(1 + r + r^2 + \cdots + r^m) = 1 - r^{m+1}$$

Simply dividing both sides of this equation by $1 - r$ and taking the limit as $r \to \infty$ gives the sum of the infinite series, since $r^{m+1} \to 0$ as $m \to \infty$ if $|r| < 1$.

Example 1.22. A coin is to be tossed repeatedly until a head occurs, at which time the tossing stops. What is the probability that an even number of tosses is required?

Solution: Since there is no upper bound that can be placed on how many tosses might be required, an infinite sample space is required to model this experiment. In fact, we can simply think of the sample space S as being the set of positive integers, where the integer 4, for example, would represent the outcome that the head occurs on the fourth toss.

How should the probabilities be assigned? In order for the head to occur on the fourth toss, tails would have been necessary on the first three tosses. The probability of getting 3 tails followed by a head is $.5^4$ when computed by any of our standard methods. In general, an integer $k \in S$ represents the outcome in which a head follows $k-1$ tails and hence has probability $.5^k$. To check consistency, observe that the sum of the probabilities of all the outcomes in S is

$$\frac{1}{2} + \frac{1}{4} + \frac{1}{8} + \frac{1}{16} + \cdots = 1$$

Factoring out 1/2 from each term here one is left with the geometric series found in Proposition 1.4 with $r = 1/2$.

The question that is asked is what is the probability of the event

$$A = \{2, 4, 6, \cdots\}$$

that is, that the head occurs on an even numbered toss. Then

$$P(A) = \left(\frac{1}{2}\right)^2 + \left(\frac{1}{2}\right)^4 + \left(\frac{1}{2}\right)^6 + \ldots = \left(\frac{1}{4}\right) + \left(\frac{1}{4}\right)^2 + \left(\frac{1}{4}\right)^3 + \ldots$$

$$= \left(\frac{1}{4}\right)\left\{1 + \left(\frac{1}{4}\right) + \left(\frac{1}{4}\right)^2 + \left(\frac{1}{4}\right)^3 + \ldots\right\} = \left(\frac{1}{4}\right)\frac{1}{1 - 1/4} = \frac{1}{3}$$

In this example we have been interested in the number of the toss on which a head will first appear. We have viewed the sample space as simply being the positive integers,

$$S = \{1, 2, 3, \cdots\}$$

For example, the integer 17 represents the outcome in which the first head appears on the seventeenth toss.

Someone might raise the question, "But isn't it also possible that a head will *never* appear?" You might be inclined to include some other element in the sample space to represent that possibility, for example, the symbol ∞. Is it correct to do this?

Actually, it doesn't matter. Since the sum of the probabilities that $1, 2, 3, \cdots$ occur is equal to 1, we would have no choice but to assign probability 0 to any additional outcome that we included in the sample space, including an outcome to represent the possibility of heads never occurring. So including it is pointless, but not necessarily incorrect.

This example involves a *countably infinite* sample space. A countably infinite set is one that can be put in one-to-one correspondence with the positive integers. Notice that the techniques used are almost identical to those used with finite sample spaces. We needed only to replace the idea of a finite sum by that of an infinite series.

In Chapter 5 we will be studying continuous models. As mentioned earlier, in continuous models the computations involve integrals rather than sums. Often these are, in fact, easier to work with. This is because the standard methods of calculus often are sufficient to evaluate integrals, whereas evaluating sums (even finite ones) can be pretty tedious at times. There is not much that can be said in this introductory chapter about continuous models, however, because the groundwork that is laid in Chapter 3 must come first.

Problems

1.1 Assume that $P(A) = .6$, $P(B) = .3$, and A and B are independent. Give the following probabilities:
(a) $P(A \cap B)$ (b) $P(A \mid B)$ (c) $P(A \cup B)$ (d) $P(A \cup B^c)$

1.2 Two devices are tested from a batch of 6 items of which 4 are good and two are defective. Find the probability that
(a) both are good.
(b) both are good given that at least one is good.

1.3 John figures the probability he will make a grade of C or better in his science course is .6 and the probability he will make a C or worse is .7. What is the probability he will make exactly a C?

1.4 Peter has 4 blue socks, 3 brown socks, and 2 white socks in his drawer. If he randomly reaches in and pulls out a pair of socks, what is the probability that they will match?

1.5 Suppose A and B are events with the properties that $P(A^c \cap B^c) = .2$, that $P(A \cap B) = .2$, and that $P(B^c) = .4$. Find $P(A)$.

1.6 A device has 5 components. Assume that each one has a probability .1 of being defective and that whether a given one is defective does not depend on the condition of any of the other components. Thus the 5 components may be viewed as an independent trials process with $n = 5$ = number of trials.
(a) For k = 0, 1, 2, 3, 4, and 5, find the probability that exactly k are good.
(b) What is the probability that at least 3 are good?

1.7 In a county with 11 incorporated towns, 7 have an approved water supply and 4 do not. An inspector intends to make a random spot check of the water supply of 2 of the towns in the county.

(a) What is the probability that both of the towns that are checked will have an approved water supply?

(b) What is the probability that neither will?

(c) What is the probability that at least one will?

1.8 A and B are going to play a tennis match. The winner must win two sets in order to win the match. (They stop playing when someone has won two sets.) In each set, the probability that A wins the set is .6.

(a) Draw a tree diagram for the tennis match.

(b) What is the probability that A wins the match?

(c) What is the probability that B wins exactly one set?

(d) What is the probability that B wins at least one set?

(e) What is the probability that A wins the match given that B wins the first set?

1.9 Assume the following to be true: Thirty percent of the members of a club play tennis. Of those who play tennis, 70% also play badminton. Of those who do not play tennis, only 10% play badminton.

(a) What is the probability that a randomly chosen person from the club plays badminton?

(b) If you know that a randomly chosen person plays badminton, what then is the probability that he or she plays tennis?

1.10 A system has five components, two of which are known to have a special marking stamped on the side. (The other three are known *not* to have the marking.) Components are checked one at a time *until two of the unmarked components have been found.*

(a) Draw a tree diagram for this experiment.

(b) What is the probability that both of the marked components are checked?

(c) What is the probability that exactly one of the marked components is checked?

(d) What is the probability that at least one marked component is checked?

1.11 In a certain population, 5% have a disease. A diagnostic test for the disease returns a positive result 90% of the time when it is used on a person who actually does have the disease. (The other 10% of the time the test fails to detect the disease.) However, when the test is used on a person who actually does not have the disease, the test returns a positive result 8% of the time.

(a) What is the probability that a randomly chosen person actually has the disease and will test positive?

(b) If the test returns a positive result for a randomly chosen person, what is the probability the person actually has the disease?

1.12 In a certain high school, 10% of the male students weigh 200 lb or more. Of those who do weigh at least 200 lb, 30% play football. Of those who weigh less than 200 lb, only 5% play football. If a randomly chosen male student is known to play football, what is the probability that he weighs at least 200 lb? Do this problem with a tree diagram and again using Bayes' theorem.

1.13 A red and a green die are rolled.

(a) Find the probability that the number appearing on red is greater than the number appearing on green.
[Hint: Let B be the event under consideration, and let $A_1, \cdots, A_6$ be the events representing occurrence of the numbers 1 through 6, respectively, on the green die. Then

$$P(B) = P(A_1)\,P(B \mid A_1) + \cdots + P(A_6)\,P(B \mid A_6)$$

This equation comes from Proposition 1.1.]

(b) Use Bayes' theorem to compute the conditional probability that the green die shows a 3 given that the number on red exceeds the number on green.

1.14 A coin is tossed and a die is rolled.

(a) Construct a natural sample space and probability measure for this experiment.

(b) Show that the events A = "a head appears on the coin" and B = "an even number appears on the die" are independent events.

1.15 Suppose S_1 and S_2 are finite sample spaces associated with two different experiments. The sample space for the combined experiment consisting of both of these experiments could be taken to be the set of all ordered pairs (w_1, w_2) where $w_1 \in S_1$ and $w_2 \in S_2$. Do you see a natural way of defining a probability measure on this sample space that will make the results of the two experiments independent of each other? (Consider this question in light of Problem 1.14.)

1.16 An electrical device has two components, A and B. Each component can be "good" or can be in a "shorted" or "open-mode" failure state. Let's assume that component A has probability .6 of being good and probability .2 of being in each of the failure states, and that component B has probability .7 of being good, .2 for "open" failure, and .1 for "shorted" failure. Furthermore, we will assume that the two components function independently.

.28 (a) What is the probability that at least one of the components is shorted?

.12 (b) What is the probability that neither is good?

.64 (c) If the two components are wired in series, what then is the probability that current can pass through?

.96 (d) If they are wired in parallel, what is the probability that current can pass through.

[Current can pass through a device if it is "good" or if it is in the "shorted" failure state.]

1.17 Among the products of a certain manufacturer, 30% are defective. If you pick 10 off the assembly line for testing, what is the probability that exactly 2 are defective? (Treat this as an independent trials process.)

1.18 A system has 100 components, 30 of which are defective. If you pick out 10 *different* components for testing, what is the probability that exactly 2 of them are defective? (Use the combinations formula.) Make sure you understand the difference in the situation described here and the situation of Problem 1.17. This is a bit subtle, for in Problem 1.17 you presumably are considering the testing to take place on 10 *different* items as well. Yet in this problem, the sampling would have to be considered to be sampling *with* replacement in order to produce the same answer as in 1.17.

(Sampling *with* replacement means that an item tested once may be tested again. In other words, each item is replaced in the original pool from which the samples are being drawn immediately after it is tested. For example, if 2 cards are drawn *with* replacement from a deck of cards, they both could be the ace of spades.)

1.19 A device is put into service on a Monday and operates seven days each week. Each day there is a 10% chance that the device will break down. (This includes the first day of operation.) The maintenance crew is not available on weekends, and so the manager hopes that the first breakdown does not occur on a weekend. What is the probability that the first breakdown *will* occur on a weekend. [Hint: Make use of the concept of a geometric series. View each day as a separate trial in an independent trials process in which you are waiting for the first success.]

1.20 Show by induction that for any collection of events $A_1, A_2, \cdots, A_n$, it is true that $P(A_1 \cup \cdots \cup A_n) \le P(A_1) + \cdots + P(A_n)$.

1.21 Do the problem given in Example 1.16 using a tree diagram.

1.22 A random digit generator outputs random digits. Each time a digit is generated, the digit is equally likely to be any one of the digits 0,1,2,...,9. How many digits must be generated in order for us to be 95% certain that at least one 7 is generated? [Hint: This means that the probability of getting at least one 7 should be greater than or equal to .95.]

1.23 Show that if a pair of events A and B is both independent and mutually exclusive, then at least one of the events must have probability zero.

1.24 Suppose that A, B, and C are independent events.

(a) Show that A and $B \cap C$ are independent.

(b) Show that A and $B \cup C$ are independent.

1.25 Draw a Venn diagram for the data in Example 1.1, using circles to correspond to the categories "low blood pressure" and "men" as opposed to "high blood pressure" and "women," as is shown in the text.

1.26 In the Venn diagram that follows, A, B, and C represent the sets of letters given in Example 1.3. For each of the 26 letters of the alphabet, place the letter in the correct one of the eight regions shown in the Venn diagram. For example, the letter x would go in the outer region (outside the 3 circles) as shown in the figure because it doesn't belong to any of the three events.

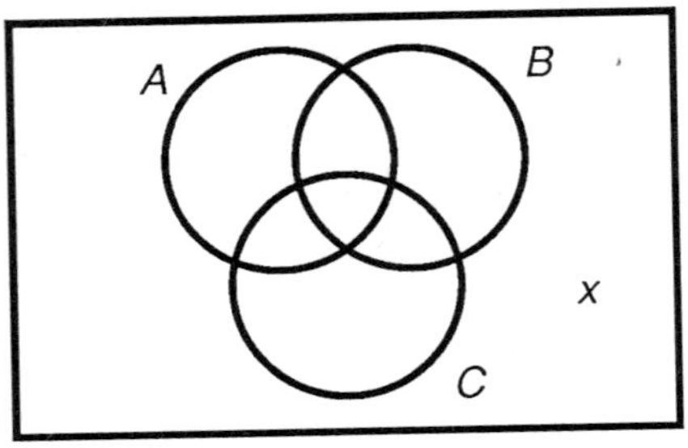

1.27 In a Venn diagram with three sets, the figure is divided into eight connected regions.

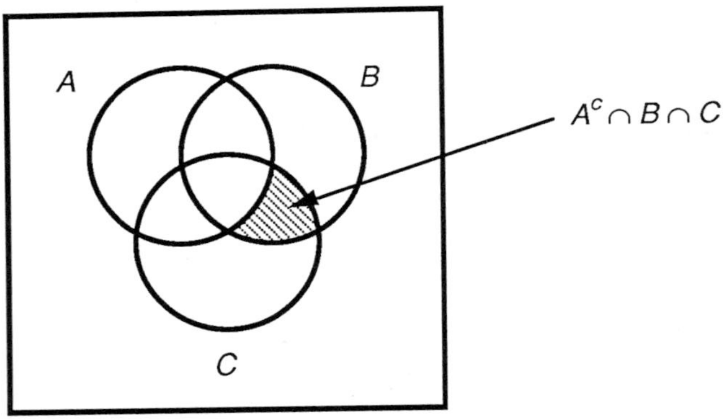

One of the eight regions is shaded in the figure, and the set that is represented by the region is described in terms of A, B, C, and the elementary set operations. Give a similar description for each of the other 7 regions in the figure.

1.28 To check whether two given events are independent, any of three different equations can be used. (See Definition 1.2.) Prove that the three equations in Definition 1.2 are equivalent. In other words, show that if one of them is true, then all of them must be true. (This is very easy. For example, you might first show that equations 1 and 3 are equivalent. By symmetry, an identical argument would show that Equations 2 and 3 are equivalent.)

1.29 Example 1.22 shows that if a coin is tossed repeatedly until a head occurs, the probability is 1/3 that the first head will occur on an even-numbered toss. Give a similar argument to show that the probability is 2/3 that the first head will occur on an odd-numbered toss.

Chapter 2: Applications

This section will give further engineering applications of the topics discussed in Chapter 1. Some such problems involve reliability of systems like circuits and communication networks. The examples given in this chapter will be elementary, but they will show the usefulness of the properties and techniques that have already been encountered.

2.1 Circuits

In Figure 2.1, each box represents a component of a circuit. It is assumed that current can flow along the lines connecting the components.

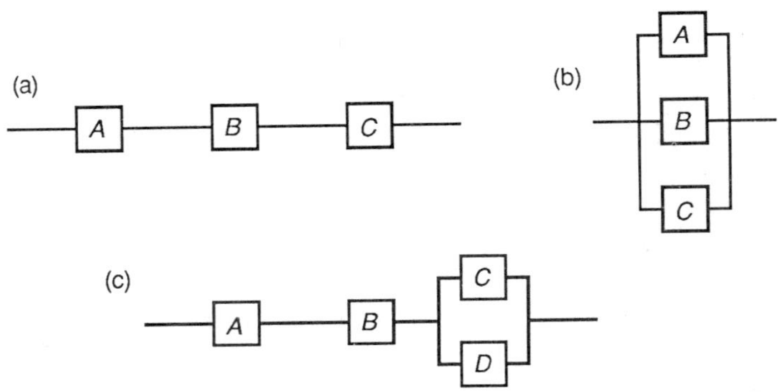

Figure 2.1 Three circuits. (a) A series circuit, (b) a parallel circuit, and (c) a series-parallel circuit.

In the circuits we are interested in determining the probability that the circuit will carry current, and we assume we know the probability that current can flow through each of the components. It is common practice to call the probability that the circuit will carry current the *reliability* of the circuit, and similarly for the individual components. (We will sometimes shorten "reliability" to "rel.") Let's denote by A, B, C, and D the events that components A, B, C, and D (respectively) in Figure 2.1 can carry current. Furthermore, *we will assume that these events are independent.* It is very important to keep this assumption constantly in mind, because not all circuits actually behave this way (though many certainly do). For example, failure of one component may cause another to overheat, thereby making failure of another component more likely.

Series Circuits

Example 2.1. The series circuit in Figure 2.1 (a) can carry current only if all components are functioning. Thus the reliability of the circuit is

$$\text{rel} = P(A \cap B \cap C) = P(A)\, P(B)\, P(C) \tag{2.1}$$

More generally, it should be clear that for any series circuit, the reliability is simply the product of the reliabilities of the individual components, assuming that the events corresponding to the components being good are independent events. (Engineers usually describe this assumption by saying that "the components fail independently.")

Parallel Circuits

Example 2.2. The parallel circuit in Figure 2.1 (b) can carry current if at least one of the components is functioning. Thus $\text{rel} = P(A \cup B \cup C)$. The easiest way to evaluate this probability is to recall De Morgan's law:

$$(A \cup B \cup C)^c = A^c \cap B^c \cap C^c$$

The fact that A, B, and C are independent guarantees that A^c, B^c, and C^c are also independent. Thus,

$$\text{rel} = 1 - P[(A \cup B \cup C)^c] = 1 - P(A^c)\, P(B^c)\, P(C^c)$$

$$= 1 - (1 - p_A)(1 - p_B)(1 - p_C) \tag{2.2}$$

where p_A, p_B, and p_C denote $P(A)$, $P(B)$, and $P(C)$, respectively.

In general, a parallel circuit with n components would have reliability

$$\text{rel} = 1 - (1 - p_1)(1 - p_2) \cdots (1 - p_n)$$

where p_i denotes the reliability of the ith component, and the components are assumed independent.

In Examples 2.1 and 2.2 we have expressed the reliability of simple series and parallel circuits in terms of the reliabilities of the individual components. If there is more reason to be interested in *failure* of the circuit than in reliability, the situation can be turned around and the *unreliability* of the circuit can be expressed in terms of the *unreliability* of the components. The *unreliability* of a circuit or component is simply the probability of failure; that is, the probability that the device cannot carry current. (See Problem 2.6.)

Series-Parallel Circuits

Example 2.3. A series-parallel circuit is a circuit built up from smaller pieces (or individual components) by hooking modules together in series or parallel. Figure 2.1c is a simple example. By visual inspection, it is apparent in this example that the circuit works only if A and B function and if either C or D does. That is,

$$\text{rel} = P[A \cap B \cap (C \cup D)]$$

Because A, B, C, and D are independent events, it follows that the events A, B, and $C \cup D$ are also independent. Thus

$$\begin{aligned} \text{rel} &= P(A)\, P(B)\, P(C \cup D) \\ &= P(A)\, P(B)\, [\, P(C) + P(D) - P(C)\, P(D)\,] \\ &= p_A p_B p_C + p_A p_B p_D - p_A p_B p_C p_D \end{aligned}$$

In general, there is no formula to represent the reliability of an arbitrary series-parallel circuit. However, a simple manual technique is always available in that the problem can always be done by a "bottom-up" approach. In other words, a group of components that form a series or parallel module can be treated as a single "super" component. The reliability of the super component is computed using the series or parallel formula, and the module is then treated as a single component, thereby creating a simpler problem to be solved. In Figure 2.1c this would amount to treating the module consisting of C and D as a super component, computing its and the new component.

2.2 Networks

From a mathematical standpoint, there is little to distinguish between a circuit and a network. Figure 2.2 illustrates this fact.

Networks often have two special nodes called the *source* and the *sink*, and this is true of the ones pictured in Figure 2.2. The links of the network are thought of as devices that may fail, and as such are analogous to the components in Figure 2.1. The nodes where the links join are analogous to the "wires" that connect the components of the circuits. It should be clear that mathematically the problem of computing the reliability of the networks in Figure 2.2 is identical to the problem of computing the reliability of the circuits in Figure 2.1. Thus the computations shown above for the circuits apply without modification to the networks.

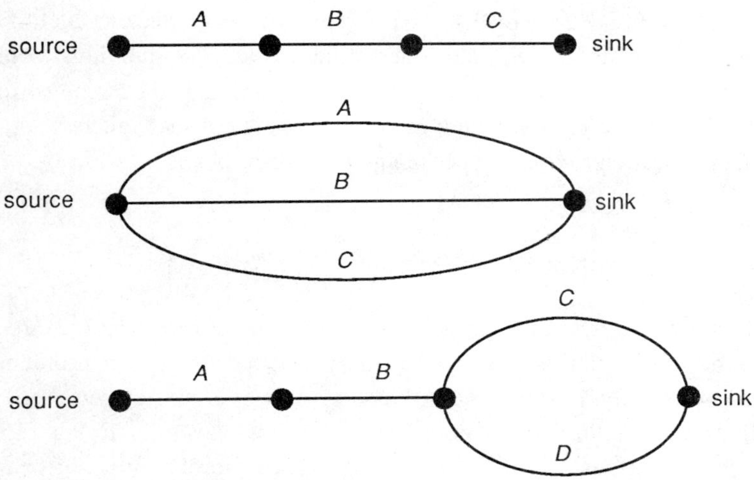

Figure 2.2 These series-parallel networks are mathematically identical to the circuits of Figure 2.1.

In the case of the circuits, it was assumed that only the components can fail; that is, we did not consider the possibility of any of the "wires" connecting the components failing. Similarly, in the case of the networks we are assuming that the nodes are perfectly reliable. (In Figure 2.2 the nodes are the black dots at the ends of the links.)

One current important class of network problems has to do with communication networks. Visualize a large group of computers or terminals (nodes) linked together by a variety of communication lines (edges). Since things like power outages or equipment failures can cause communication links to fail, there is a clear need to study the reliability of such systems. It is also possible to study systems in which both the vertices and the edges can fail. We did not allow for that possibility in treating Figures 2.1 and 2.2 above.

Graphs

Both networks and circuits, when viewed from a purely mathematical point of view, are instances of what are called *graphs*. A graph consists of "edges" and "vertices." When drawn, the edges are usually lines or curves, and the vertices are the junction points where edges meet. Thus graphs are visually similar to the networks of Figure 2.2. To identify a circuit with a graph, one identifies the components of the circuit with the edges of the graph. The individual connected clumps of "connecting wires" that connect the components in the circuit correspond to the vertices of the graph. Graph theory is a branch of mathematics which has important applications to many types of engineering problems.

Networks That Are Not Series-Parallel

Example 2.4. The nice thing about series-parallel networks (or circuits) is that reliability calculations can be manually reduced to simpler problems in a straightforward way. More general kinds of networks do not necessarily lend themselves to such direct treatment. Figure 2.3 illustrates this. In this figure we assume that the edges A, B, C, D, and E have known probabilities of being in a working condition and that these events are independent.

If w and y are the source and the sink, then the network is series-parallel and the reliability is the probability of the event $(A \cap B) \cup E \cup (C \cap D)$.

If x and z are source and sink, then the problem becomes substantially more difficult. It is no longer possible to merge series or parallel components. There are a variety of approaches, however, that can be used. Perhaps the simplest is to "partition" the problem depending upon whether link E is working or not. To see what this means, let's let G represent the event that a communication route from x to z is available. Then

$$P(G) = P(E)\, P(G \mid E) + P(E^c)\, P(G \mid E^c)$$

(In this equation the event E denotes the event that link E is functioning.) This is Proposition 1.1, the multiplicative law. Since $P(E)$ and $P(E^c)$ are assumed known, we yet need to know how to evaluate $P(G \mid E)$ and $P(G \mid E^c)$.

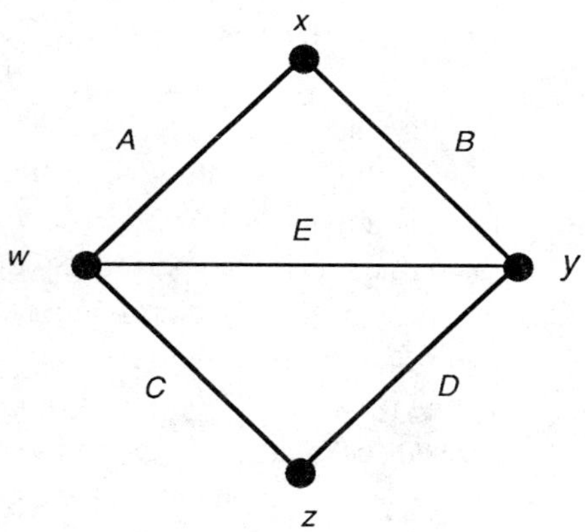

Figure 2.3 If w and y are viewed as the source and the sink, this is a series-parallel network. If x and z are source and sink, the network is not series-parallel.

But if edge E is good, we might as well think of w and y as a single vertex (since we know they are connected). In order to get to this single vertex from x and from z, at least one of the edges A and B has to be good and at least one of C and D must be good. This event is $(A \cup B) \cap (C \cup D)$. So,

$$\begin{aligned} P(G \mid E) &= P[\,(A \cup B) \cap (C \cup D)\,] = P(A \cup B) \times P(C \cup D) \\ &= (p_A + p_B - p_A p_B) \times (p_C + p_D - p_C p_D) \end{aligned}$$

On the other hand, if edge E is bad, then we can simply remove it from the picture. In this case, a path from x to z will be open only if A and C are good or if B and D are good. This event is $(A \cap C) \cup (B \cap D)$. Thus,

$$\begin{aligned} P(G \mid E^c) &= P[(A \cap C) \cup (B \cap D)] \\ &= P(A \cap C) + P(B \cap D) - P(A \cap B \cap C \cap D) \end{aligned}$$

Each of these terms can now be expanded as a product because of the assumed independence.

There is a conceptual point here that is more important than the nitty-gritty details. It is that by "pivoting" on the edge E, that is, by splitting the problem into two cases depending upon whether edge E is good or bad, the problem can be split into two simpler problems. This idea plays a key role even in very sophisticated techniques and in current research.

Even very important contemporary problems needn't involve highly abstract concepts. The network reliability problems introduced above illustrate this fact. Suppose that a group of telephone users are hooked together by a network of telephone lines of known reliabilities and that whether a given line is down is independent of whether any of the other lines are down. It is hypothetically possible to determine the probability that all individuals can communicate with each other or that any particular subgroup can communicate with each other. This calculation can be done using a variety of the techniques developed in Chapter 1. Some of the techniques can be refined into quite powerful tools, while others have to be discarded as too crude. All methods of determining network reliability are very "inefficient" algorithms in that the time required for the calculation grows exponentially with the size of the network. The amount of time or the number of computations required to perform a calculation is referred to as *the computational complexity* of the method. The development of computer algorithms for performing such analyses is an important area of contemporary research.

2.3 Case Studies: Two Network Reliability Problems

Network reliability problems provide a useful illustration of several of the basic notions of probability and conditional probability that were discussed in Chapter 1. In performing parallel and series reductions in a circuit or a network, we use fundamental properties of union and intersection since a series reduction corresponds to an intersection of two events and parallel reduction corresponds to a union of two events. The idea of independence is involved at every step of the way.

In this section we will examine how using the concept of conditional probability can help with more complicated systems than series-parallel configurations. As mentioned earlier, problems such as this are in the forefront of present-day research because of the increasing importance of communication and transportation hookups of one kind or another around the world.

Example 2.5. In Figure 2.4, the five circles in the network G_0 at the top represent five people linked together via the seven communication lines shown in the picture as edges $e_1, \cdots, e_7$. We will assume that the reliability of each of the links is known; that is, that the probability the link is working is known. Furthermore, let's assume that the links are independent. As a convenient notation, we denote by p_k the probability that link e_k is good.

The problem we are going to investigate is the problem of determining the reliability of communication between the two people represented by the black nodes. What are some possible approaches that could be used?

Tree Diagrams

Here we are looking at seven links, each of which is going to be either good or bad. To represent all states of this system in a tree diagram, we would need to have seven layers of branching in the tree, one layer for each line. Since the number of outcomes shown in the tree will double for each layer, we will wind up with $2^7 = 128$ outcomes in the tree.

In reality it isn't necessary to draw the entire tree in order to do this problem. For instance, suppose we were to start out by considering edge e_3 first and edge e_2 second as we start building the tree. If both these links are bad, then certainly it is going to be impossible for the two black nodes to communicate. Therefore, there isn't any point in continuing to draw the part of the tree that would continue on from the assumption that that these two links have both failed, if all we are interested in getting from our tree diagram is the probability that the black nodes can communicate. Even so, this problem clearly taxes the methodology of a tree diagram to its limit. While it may be (barely) feasible to do this problem with a tree diagram, given a very large sheet of paper, it wouldn't be much fun.

Finding All Possible Communication Paths

There are clearly many routes that could be used to send a message from one black node to the other. A few are $e_3\, e_4\, e_5$ and $e_2\, e_6$ and $e_3\, e_1\, e_7\, e_5$. For any one of these possible paths, we can easily compute the probability of the path being available because of the independence we are assuming for the links. For example, the probability that the path $e_3 e_4 e_5$ is a working path is simply the

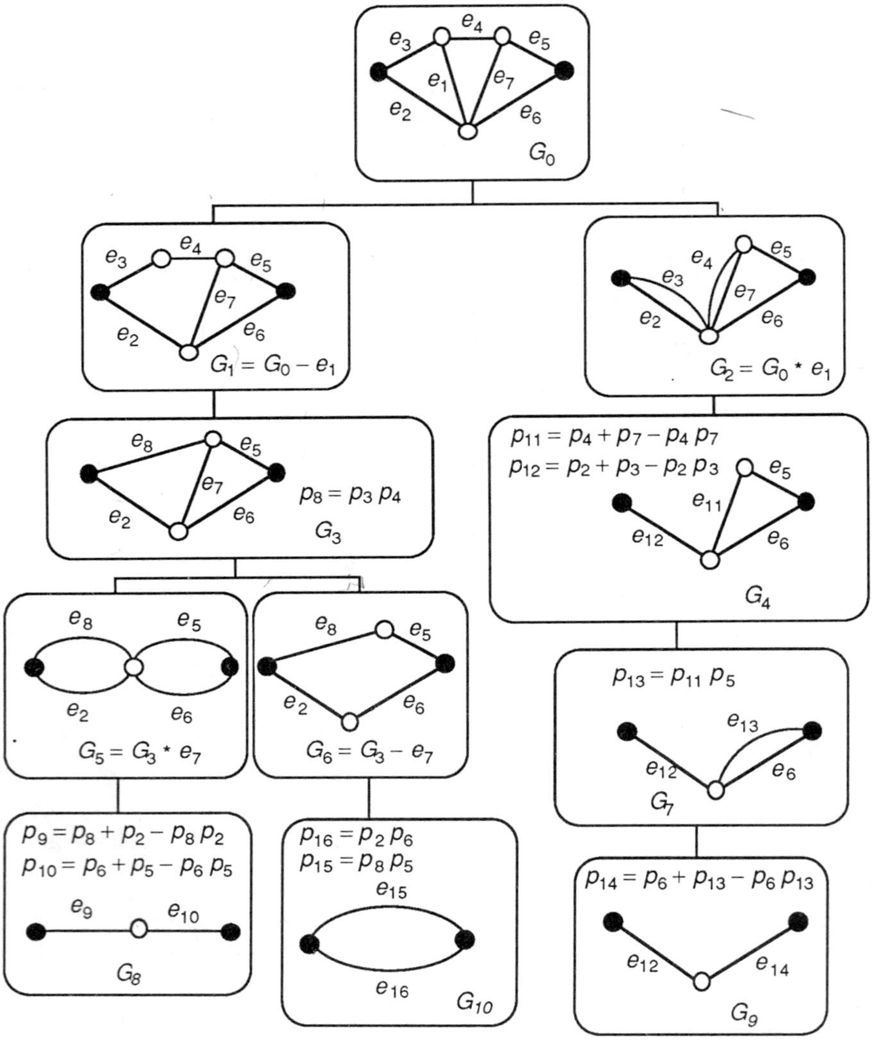

Figure 2.4 Network reliability using recursion and conditional probabilities.

product $p_3 \times p_4 \times p_5$, since it depends on all three of the three edges involved being good. Therefore, we could list all the possible paths and compute the probability for each path. The problem with this approach, however, is that the different paths aren't mutually exclusive. For example, consider the two paths $e_3e_4e_5$ and $e_3e_1e_7e_5$. Let's think of event A as being the event that the first of these two paths is good and event B as the event that the second is good. Clearly A and B aren't mutually exclusive, because it's quite possible that both paths might be good, and so we can't compute $P(A \cup B)$ just by adding $P(A)$ and $P(B)$. The best we can do is

$$\begin{aligned} P(A \cup B) &= P(A) + P(B) - P(A \cap B) \\ &= p_3 p_4 p_5 + p_3 p_1 p_7 p_5 - p_3 p_4 p_5 p_1 p_7 \end{aligned} \tag{2.3}$$

This computation is easy, since $A \cap B$, the event that both paths are good, depends only upon e_3, e_4, e_5, e_1, and e_7 being good, and the probability of this happening is given by $p_3 p_4 p_5 p_1 p_7$.

So if there were only two possible paths connecting the black nodes, the computation we have just described would do the trick. The problem is that there are many such paths, not just two. Let's think of n as representing the number of paths between the black nodes and $A_1, \cdots, A_n$ as being the events "path 1 is good," "path 2 is good," and so forth. Since what is needed in order for communication to be possible is for *at least one* of the paths to be good, the task is to compute the probability $P(A_1 \cup \cdots \cup A_n)$.

There is a standard way to do this using what is sometimes called the *inclusion-exclusion principle.* In fact, Equation 2.3 above is the special case of this principle when $n = 2$. For $n = 3$ the inclusion-exclusion principle says that

$$\begin{aligned} P(A \cup B \cup C) = \; & P(A) + P(B) + P(C) \\ & - P(A \cap B) - P(A \cap C) - P(B \cap C) \\ & + P(A \cap B \cap C) \end{aligned}$$

For the union $A_1 \cup \cdots \cup A_n$, the inclusion-exclusion principle says that the probability $P(A_1 \cup \cdots \cup A_n)$ can be computed by adding the probabilities of each of the individual events, then subtracting off the probability of the intersection of all possible pairs of the events, then adding on the probabilities of all intersections of three events, subtracting off the probabilities of all intersections of four, and so forth. While it's not terribly difficult to see why this principle is correct (in fact, it can be derived from Equation 2.3 using mathematical induction), it clearly isn't a recipe that's going to be very pleasant to use if n is large.

Network Reductions Using Conditional Probabilities

In earlier examples we have seen how to use series and parallel simplifications to determine the reliability of a series-parallel circuit or network. Even in networks that are not of a totally series-parallel structure, it's still often possible to make such simplifications. The problem with the network G_0 in Figure 2.4, however, is that there aren't *any* edges in the network that can be reduced this way, and so there is no apparent way to get started.

A useful starting point is to focus on a single edge (any one will do) and consider the two possibilities: Either the link is good or it is bad. We can use the concept of conditional probability as embodied in the multiplicative law of Proposition 1.1.

To make things specific, let's fix our attention on edge e_1. Denote by E the event that this edge is good so E^c is the event that edge e_1 is bad. We also need a name for the event whose probability we are trying to compute, so let's designate A as the name of the event that the two black nodes in network G_0 can communicate. Proposition 1.1 says that we can determine the probability of this event by

$$P(A) = P(E)\,P(A \mid E) + P(E^c)\,P(A \mid E^c)$$

The usefulness of this little formula lies in the fact that there is a very natural interpretation to each of the conditional probabilities contained in it. The conditional probability $P(A \mid E)$ is the probability that communication is possible given that the link e_1 is good. If e_1 is good,however, then we know for certain that the two nodes at the ends of e_1 will be able to communicate, and so any message that can get to one can automatically get to the other. Why not then think of them as a single node? In other words, take e_1 out of the picture altogether and "collapse" the two nodes at the ends of e_1 into a single logical node, because being able to get to one of them is equivalent logically to being able to get to the other one given that e_1 is good. What this observation boils down to is the fact that the conditional probability $P(A \mid E)$ is the same as the *unconditional* probability that the two black nodes can communicate in network G_2 of Figure 2.4. On the other hand, the conditional probability $P(A \mid E^c)$ has an even simpler interpretation. For if e_1 is bad, we might as well remove it from the picture altogether. The conditional probability $P(A \mid E^c)$ is the same as the probability that the two black nodes could communicate if e_1 weren't there at all, and this is what is shown in network G_1 in the figure. The conditional probability $P(A \mid E^c)$ is the same as the *unconditional* probability that the black nodes can communicate in the network G_1.

It is common to refer to G_1 as the network (or graph) obtained from G_0 by *deleting* the edge e_1 and to refer to G_2 as the network obtained from G_0 by *contracting* the edge e_1. Notationally this is often written as $G_1 = G_0 - e_1$ and $G_2 = G_0 * e_1$, as is indicated in the figure.

You may not have noticed, but a very important thing has just happened. We have described how to solve our original problem by expressing the solution in terms of two simpler problems. Specifically, we have passed from having to deal with network G_0 to having to deal with networks G_1 and G_2. Furthermore, each of these networks can be further simplified by series and parallel reductions. For example, e_3 and e_4 in G_1 are series edges, and G_3 shows them being replaced by a single edge e_8 whose reliability is given by $p_8 = p_3 \times p_4$. Similarly, G_2 has e_2 and e_3 as parallel edges, as well as e_4 and e_7, and G_4 shows these pairs each being replaced by a single edge with the appropriate probability. G_7 and G_9 continue this process, and in fact G_2 can be completely solved by series and parallel reductions.

In solving G_1, however, we are stuck again when we get to G_3, because G_3 has no series or parallel edges. For that reason we have to repeat the trick we used to start with, that is, to pick another edge and to delete and contract it leading to two further problems. In the figure, the edge chosen is e_7 (the particular choice of edge is not critical), and deleting and contracting e_7 leads to the two networks G_6 and G_5, respectively. G_5 is then reduced to G_8 and G_6 to G_{10}.

All known methods for doing problems such as this one require a number of computational steps that grows exponentially with the size of the network. Reduction methods similar to what is shown in Figure 2.4, however, work about as well as any known method. In fact, it is possible to develop algorithms using such reduction methods and to run the algorithms on a microcomputer.

Figure 2.5 shows a much more complicated network than the one we have been considering. In fact it is based on a computer network that existed in the 1970s with source in California and sink in New England. The probability that the two black nodes can communicate in the network of Figure 2.5 can be calculated on a microcomputer in less than a minute using an algorithm that implements the ideas of Figure 2.4. Since the network has 22 links, a complete tree diagram for all states of the network would show $2^{22} = 4{,}194{,}304$ outcomes.

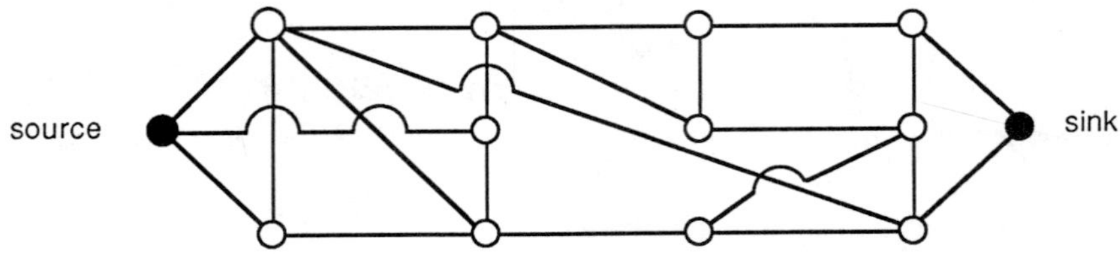

Figure 2.5 The network reliability problem is to determine the probability that communication is possible between the source and the sink.

The lengthy example we have been treating is an instance of the *two-terminal*, or *source-to-sink*, communication problem, since we are interested in determining the probability that two given members of the network can communicate. Another variation is the *all-terminal* problem in which we wish to determine the probability that *all* members can communicate with each other.

Example 2.6 (All-terminal reliability). The network G_1 that is shown in Figure 2.6 has five nodes, all of which must be able to communicate with each other. Here there is no concept of a source and a sink, but the concept of series and parallel edges is still meaningful. For example, the edges e_1 and e_2 are in series in the network on the left. The network G_2 on the right shows these two edges being replaced by a single edge e_3. Let's examine the relationship between the two networks G_1 and G_2 in the figure.

First, think about the node v_1 at the top of network G_1, the node where edges e_1 and e_2 meet. Such a node is called a *degree-two node* or a *degree-two vertex* because there are exactly two edges coming into the node. In order for this node to be able to communicate with any of the others, at least one of the edges e_1 and e_2 must be good. Let's introduce some names for events so as to be able to write this down as an equation. We will let A denote the event that all nodes can communicate in the network G_1 and C denote the event that at least one of the edges e_1 and e_2 is good. Then

$$\begin{aligned} P(A) &= P(C)\,P(A \mid C) + P(C^c)\,P(A \mid C^c) \\ &= P(C)\,P(A \mid C) \end{aligned}$$

The last expression comes from the fact that $P(A \mid C^c) = 0$ in the first equation.

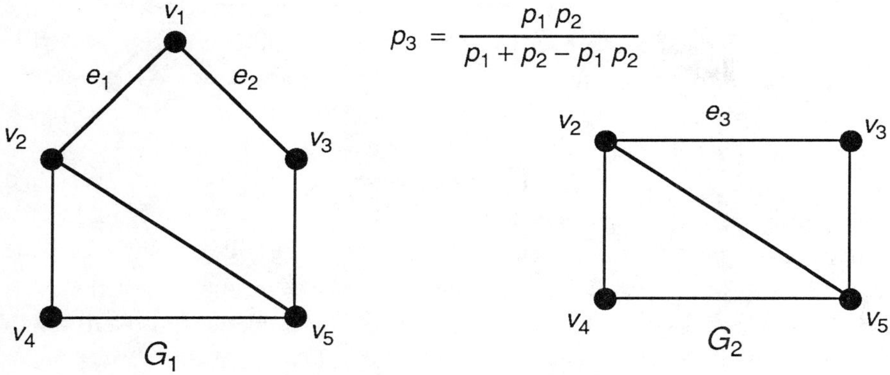

$$p_3 = \frac{p_1 p_2}{p_1 + p_2 - p_1 p_2}$$

$$\text{rel}(G_1) = (p_1 + p_2 - p_1 p_2)\ \text{rel}(G_2)$$

Figure 2.6 An all-terminal reliability problem. All nodes in network G_1 must be able to communicate with all other nodes in G_1.

Clearly $P(C) = p_1 + p_2 - p_1 p_2$, where p_1 and p_2 are the reliabilities of e_1 and e_2, respectively. What about the term $P(A \mid C)$? Given that C occurs, we know that the node v_1 is not cut off from all other nodes. Furthermore, it is possible that v_2 and v_3 can communicate through v_1 if *both* e_1 and e_2 are good. Here's where things become a bit subtle. If we knew for certain that *exactly* one of the edges e_1 and e_2 were good, we could simply remove e_1, e_2, and v_1 from the network, and the reliability of the reduced network would be the same as the original. However, since both might be good, what we need to do is replace them by link e_3 in the reduced network and to assign the proper probability to e_3. The reason for the presence of e_3 is to take care of the possibility that e_1 and e_2 can provide a communication route from v_2 to v_3. The probability of this is the probability that both e_1 and e_2 are good, which is $p_1 p_2$. Given the information that at least one of them is good, the conditional probability that both are good is the probability p_3 shown in the figure.

The relationship between the two networks is described by the equation at the bottom of the figure. The reliability of G_2 is the *conditional probability* that G_1 is good given that at least one of the two original edges e_1 and e_2 is good. The important point though is that the equation

$$\text{rel}(G_1) = (p_1 + p_2 - p_1 p_2)\ \text{rel}(G_2)$$

enables us to describe the solution to our original problem in terms of the solution to a simpler problem. In other words, what is being demonstrated is a simple algorithm for reducing two edges in series when the problem being solved is determination of the probability that *all* terminals can communicate.

Parallel reductions are no different in this example than in previous examples. So the reliability of the network G_1 in Figure 2.6 can be determined by continuing this process. For example, the degree-two vertex v_3 in G_2 can be removed in a similar fashion, and this will create parallel edges between v_2 and v_5. After the parallel edges are reduced to a single edge, we are left with a triangle of three degree-two vertices. Any of them can be removed in the same way that v_1 was, and that will leave us with an elementary network consisting of two parallel edges. By collecting all the computations, this gives us the reliability of network G_1 as some multiplying factor times the reliability of this elementary network.

So What?

Examples 2.5 and 2.6 have been used to show how solutions to important contemporary problems can be formulated in such a way that the probability concepts involved are basically elementary. Some other very important problem-solving techniques have come up also. In fact, it might be worthwhile to list them.

1. A conditional probability described in one sample space is often simplified by viewing it as an unconditional probability in another sample space. For example, if a red and a green die are tossed, the conditional probability that the sum is 8 given that the red die shows a 5 is the same as the *unconditional* probability that the green die shows a 3.

   In Examples 2.5 and 2.6, conditional probabilities involving networks have been logically replaced by unconditional probabilities involving smaller networks.

2. The concept of *recursion* is very important in problem solving, and it has been illustrated here. To illustrate recursion, think about the simple function $f(n) = n!$ for non-negative integers n. It is possible to define f by saying

$$f(n) = \begin{cases} n \times f(n-1) & \text{if } n > 0 \\ 1 \quad \text{if } n = 0 \end{cases}$$

While this definition looks redundant, it actually is quite usable. For example, to compute $f(5)$ it says you should first find $f(4)$, but to do that you have to evaluate $f(3)$, and so forth, until you get down to $f(0)$, at which time you can wrap up all your calculations.

The equation $\text{rel}(G_1) = (p_1 + p_2 - p_1 p_2)\ \text{rel}(G_2)$ in Figure 2.6 is similarly recursive in that it tells how to solve the all-terminal reliability problem for network G_1 in terms of the all-terminal reliability problem for network G_2, which is simpler.

3. Different kinds of physical systems can be modeled with the same mathematical model. Figures 2.1 and 2.2 illustrate the equivalence of some small circuit and network problems. Figures 2.4 and 2.5 are drawn as networks rather than as circuits, but they could provide mathematical models for either. For example, each of the edges in the network could be interpreted as being a resistor, and the source and sink could be incoming and outgoing wires to be connected to the terminals of a battery. The white nodes are then junction points where the wires coming out of the resistors are soldered together. A resistor is considered good if current can pass through it.

   The reliability problem we discussed in Example 2.5 then would be the determination of the probability that current will flow in the circuit when the source and sink are connected to the battery terminals.

2.4 Fault Trees

Frequently it is highly desirable to know the likelihood that a complex system will fail. One tool that is sometimes used to analyze complex systems is a *fault tree*. Figure 2.7 shows a hypothetical fault tree that has been constructed to analyze a pressure tank. The event that is of interest (and which is represented by the top node in the fault tree) is the event in which the tank explodes.

There are three kinds of nodes that appear in the fault tree. The events represented by circles are the "leaves" in the tree and represent basic events that may contribute to failure of the tank. The AND gates and OR gates represent logical intersections or unions. The very top node in the tree (which represents the event that the tank explodes) is an AND gate. This means that the tank explodes if the events corresponding to nodes 2 and 3 occur, that is, if there is high pressure and valve failure. Similarly, the valve fails (the event corresponding to node 3) if the valve is defective or if the electricity supply is interrupted.

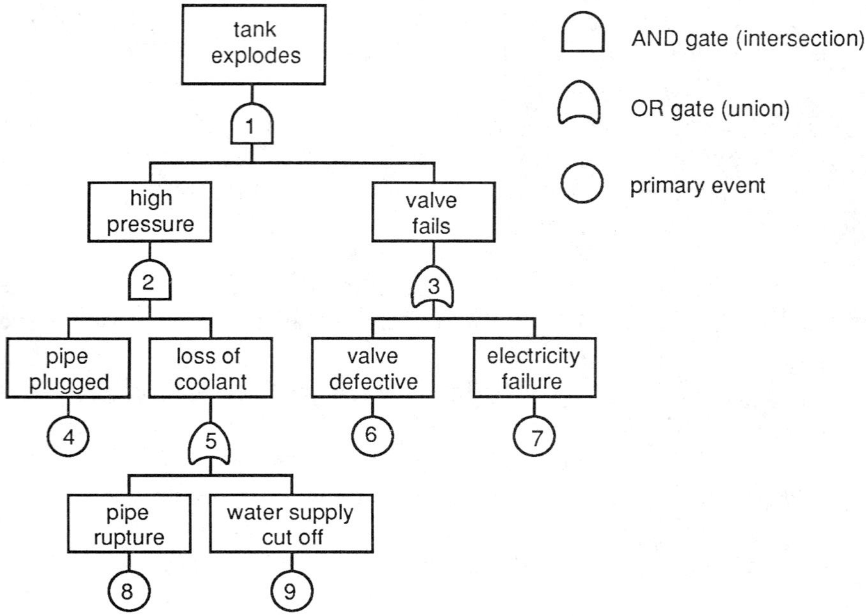

Figure 2.7 Fault tree for failure of a pressure tank.

Sometimes it is possible to construct the fault tree in such a way that the primary inputs will be independent events. (Some of the most serious defects in safety analyses, however, have centered around the assumption of independence of events that, in fact, were later found not to be independent. So independence assumptions should be made with a great deal of caution. For instance, in Figure 2.7 if water is supplied by an electric pump, then nodes 9 and 7 will not be independent unless the electricity supplying the pump comes from a different source than that described in node 7.)

If the primary inputs *are* assumed to be independent, then it is easy to compute the probability of the top event by a simple bottom-up hand calculation.

To make things concrete, let's introduce some hypothetical probabilities:

	Event	Probability
node 4	pipe plugged	.1
node 6	valve defective	.01

node 7	electricity failure	.05
node 8	pipe rupture	.05
node 9	water supply cut off	.02

Then,

$$P(\text{node } 5) = .05 + .02 - (.05)(.02) = .069$$
$$P(\text{node } 2) = (.1)(.069) = .0069$$
$$P(\text{node } 3) = .01 + .05 - (.01)(.05) = .0595$$
$$P(\text{node } 1) = (.0069)(.0595) = .00041055$$

Normally fault trees require much more subtle approaches than is shown here. Suppose, for instance, that there were more than one way in which electricity failure could contribute to failure of the tank. In this case we would have the same primary input (node 7) appearing at more than one location in the fault tree. This means that the kind of bottom-up approach used here would not work because subtrees beneath certain nodes would fail to be independent.

Fault tree analysis was developed primarily within the aerospace and nuclear industries during the 1960s and 1970s. Better ways of calculating or estimating the probability of the top event for very large and complex trees (sometimes having hundreds or even thousands of nodes) are still being sought.

Example 2.7. Draw a fault tree for the circuit of Figure 2.1c. The top event of the fault tree should represent the event that the circuit can transmit circuit.

Solution: There is not a unique way of constructing such a fault tree. Either of the following trees would be correct, and, in fact, it should be easy to convince yourself that these two fault trees are logically equivalent.

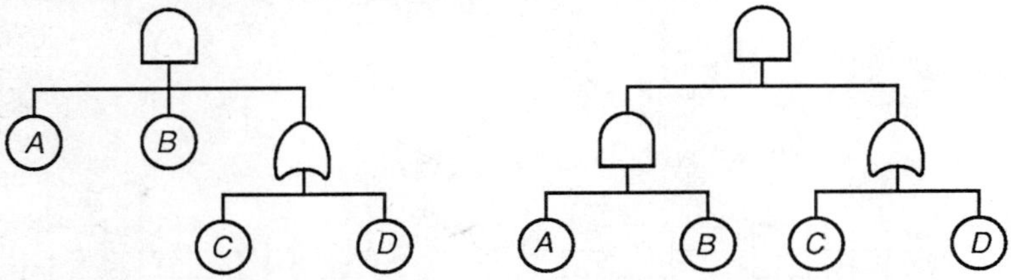

Notice that if we wish the top event to represent failure of the circuit, then a different fault tree is called for. Such a tree is the "dual" of a tree that represents "non-failure" of the circuit. Simply convert all the AND gates to OR gates and all the OR gates to AND gates. For example, here are the duals of the two fault trees above:

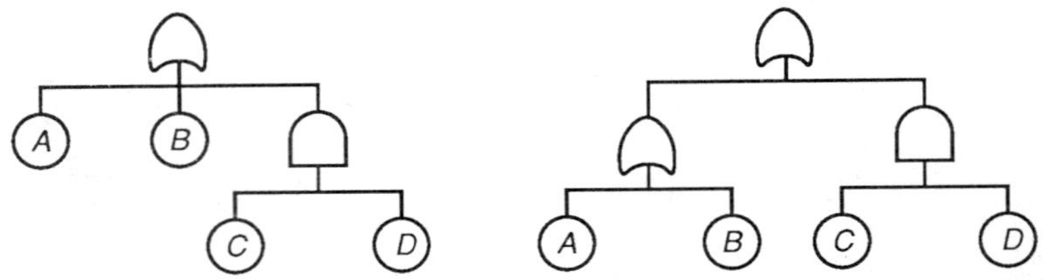

In either of these fault trees the top event represents failure of the circuit if we now interpret the primary inputs A, B, C, and D to represent failures of the respective components.

Problems

2.1 (a) What is the reliability of 5 items in parallel if each has reliability .6?

(b) What is the reliability of 5 items in series if each has reliability .6?

2.2 Find the reliability of the circuits shown below. The reliability of each component is shown with the component.

(a)

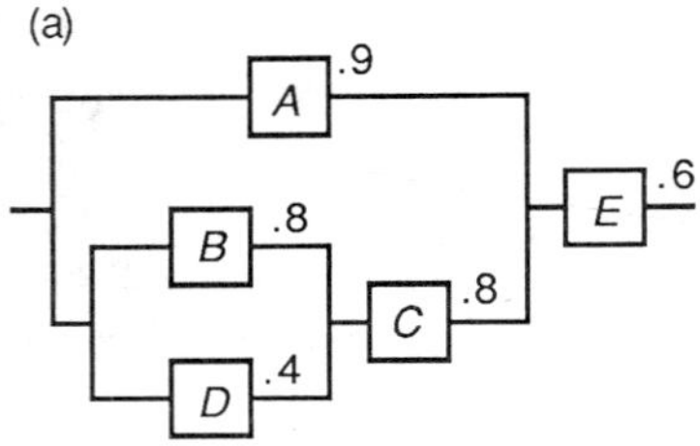

(b)

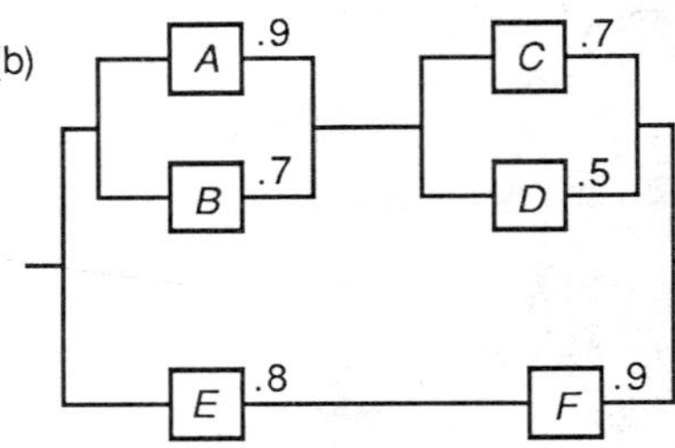

2.3 Find the reliability of the circuits shown below. The reliability of each component is shown with the component.

(a)

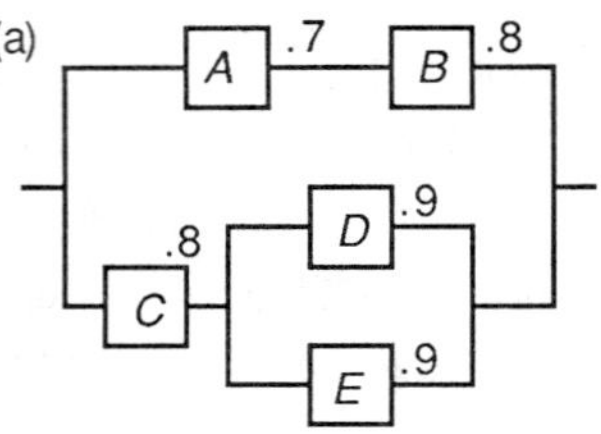

(b)

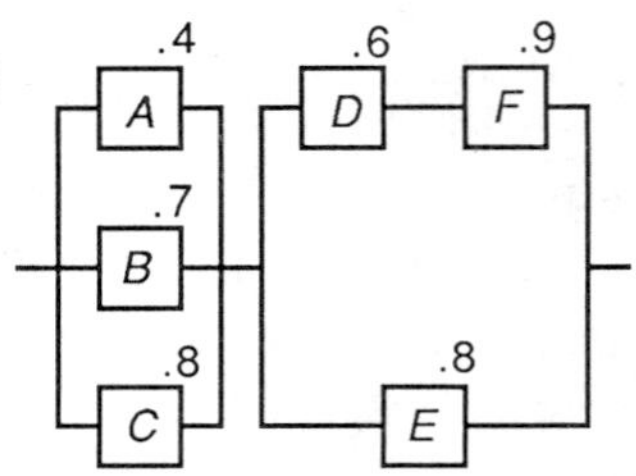

2.4 The following figure is a highway network. Consider each of the 5 links to function independently of the others.

(a) Assume that for each link the probability is .9 that the link is open. What is the probability that you can get from "start" to "finish"? (As an instructive exercise and a check of your answer, it might be worthwhile to list a sample space for this network. Since there are 5 links (edges), there are $2^5 = 32$ possible states (outcomes). One such state, for example, would be to have link A be good, B good, C bad, D bad, and E good. By the assumed independence of the links, the probability of this state

$$= P(A \cap B \cap C^c \cap D^c \cap E)$$
$$= P(A)\,P(B)\,P(C^c)\,P(D^c)\,P(E)$$

The desired probability then can be obtained by just adding up the probabilities of all of the 32 states for which it is possible to get from "start" to "finish.")

(b) Now drop the assumption that each link has reliability .9 and simply write the reliability of the network in terms of the reliabilities of the 5 links.

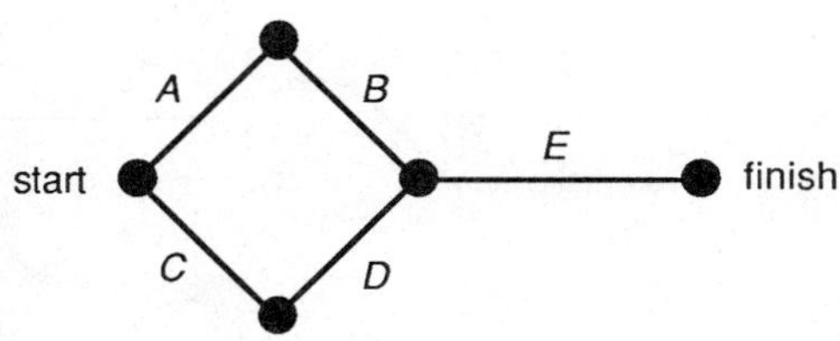

2.5 Find the probability that a route is open from the source to the sink in the network that follows. The reliability of each edge is listed alongside the edge.

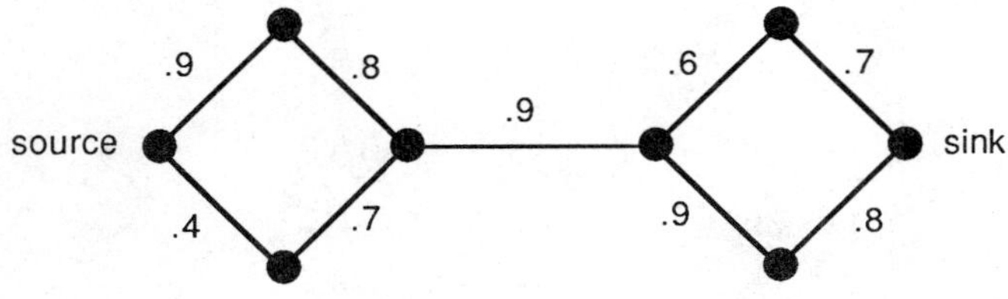

2.6 Examples 2.1 and 2.2 show how the reliability of simple series and parallel circuits are expressed in terms of the reliabilities of the individual components. Show how the *failure* probability (the *unreliability*) for the circuits in Figure 2.1(a, b) can be expressed in terms of the *failure* probabilities of the individual components. Let q_A, q_B, and q_C denote the failure probabilities of components A, B, and C, respectively.

2.7 Show how the unreliability of the circuit in Figure 2.1(c) can be expressed in terms of the unreliabilities of the failure probabilities of the individual components. (If you have already done Problem 2.6, the meaning of this should be clear.)

2.8 The circuit in the following picture shows a battery, a light, and two switches for redundancy. The two switches are operated by different people, and for each person there is a probability of .9 that the person will remember to turn on the switch. The battery and the light have reliability .99. Assuming that the battery, the light, and the two people all function independently, what is the probability that the light will actually turn on?

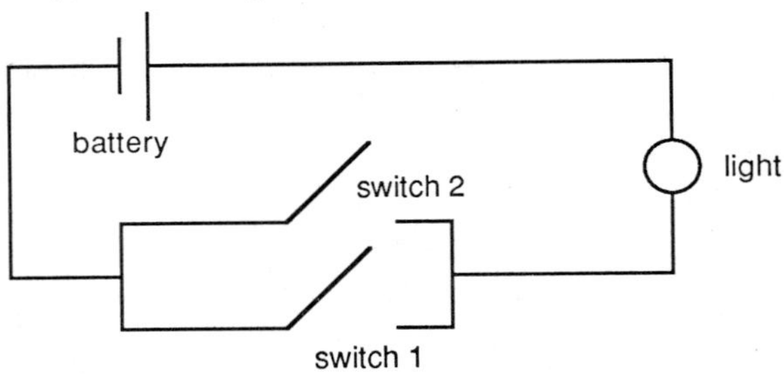

2.9 In the figure below, the probability that the battery is good is p_4, the probability that the light bulb is good is p_5, and the probabilities that the resistors R_1, R_2, and R_3 are good are p_1, p_2, and p_3, respectively.

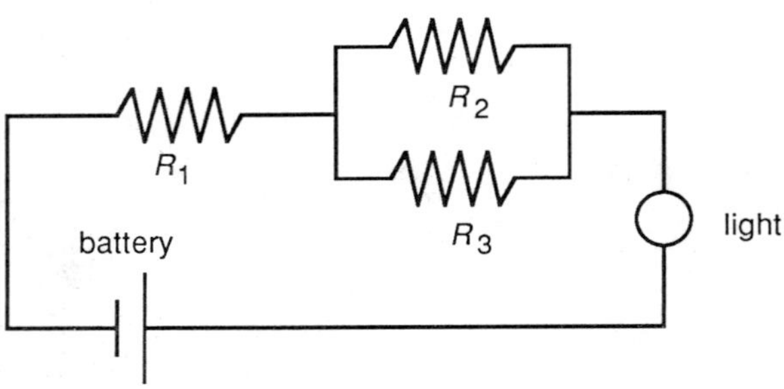

Assuming that the 5 components function independently, what is the probability that the bulb will light? (For a resistor to be "good" here means that current passes through it, and it is assumed that the bulb will light if the bulb is good and current passes through the bulb.)

2.10 A complex system requires that a certain function be performed with reliability .9999. Devices that perform the function have reliability only .8, and so it is necessary to build in redundancy. Several of the devices are to be used, and if at least one works then the function will be performed. Furthermore, we are willing to assume that the redundant devices function independently of each other. How many of the devices must be installed in order that the system will have the desired reliability?

2.11 This problem is a simple example of the all-terminal reliability problem. Three computers are all linked together over telephone lines. (If you like you can think of this as a network that is a simple triangle with the edges being the phone lines and the vertices the computers.) We'll assume that the computers are completely reliable (an unreasonable assumption) but that each of the phone lines has probability .9 of being good. Also assume the phone lines are independent. What is the probability that each of the computers will be able to communicate with the other two? [Hint: This situation is simple enough to be treated via basic principles without recourse to the ideas presented in Example 2.6.]

2.12 Suppose A, B, and C are events having the properties that

$$P(A \cap B \mid C) = P(A \mid C)\, P(B \mid C) \text{ and}$$
$$P(A \cap B \mid C^c) = P(A \mid C^c)\, P(B \mid C^c)$$

It would make sense then to call A and B *conditionally independent* relative to C and C^c. For either C does or does not occur, and the two equations say that in either case A and B are independent relative to the conditional probability measure induced by the information that C does or does not occur.

(a) Box 1 contains 2 red and 1 white items and box 2 contains 1 red and 2 white items. One of the boxes is randomly selected (each with probability 1/2), and two items are chosen from the box one at a time and without replacement. Let A be the event that the first item drawn is white, let B be the event that the two items drawn are of the same color, and let C be the event that the items are drawn from box 1. Show that A and B are independent events, but not conditionally independent relative to C and C^c in the above sense. (This is easy to check with a tree diagram, but you should also try to understand intuitively why A and B are independent and why they cease to be independent once we know what box we are drawing from.)

(b) Suppose again that box 1 contains 2 red and 1 white items and box 2 contains 1 red and 2 white items. Again randomly choose one of the boxes and draw two items, only this time replace the first item before drawing the second. Let A be the event that the first item is red, let B denote the event that the second item is red, and let C be the event that the items are being drawn from box 1. Show (via a tree diagram) that A and B are conditionally independent relative to C and C^c, but that A and B are not independent. This says that A and B become independent once we know which box we are drawing from, but in the absence of this knowledge, A and B are dependent.

* 2.13 Fill in the details of the calculation in Example 2.5. Suppose that each of the links shown in the original network G_0 has reliability .9. Determine then, with the assumption that the links function independently, what the probability is that the source and sink shown as the black nodes in network G_0 will be able to communicate.

2.14 Complete Example 2.6. Assume that all links have reliability .9 and that the links function independently. What is the probability that every node will be able to communicate with every other node?

2.15 Draw a fault tree for each of the circuits of Problems 2.2 and 2.3. First construct a fault tree in which the top event is the event that the circuit is in working order, and then give the dual fault tree in which the top event represents the failure of the circuit. (See Example 2.7.)

Chapter 3: Random Variables

The study of calculus begins with an investigation of properties of the real number system, but it does not progress very far before the concept of a function must be introduced. Similarly, in the study of probability we start with sample spaces and probability measures, but modeling real problems leads very quickly to the need to use functions as a way of describing measurements being made or processes being observed. Generally these functions are real-valued (just like in calculus), but they measure some relevant information about an experiment or a process having probabilities associated with the different outcomes or states of the process. In essence this simply says that the object we are usually interested in is a function defined on some sample space where there is a probability measure.

Definition 3.1: A real-valued function defined on a sample space S is called a *random variable.*

One of the essential things that random variables do for us is to enable us to avoid dealing with unnecessarily complicated sample spaces. Suppose, for example, you were doing quality-control analysis of a particular product such as a line of microcomputers. Some quantities you might be interested in could include number of repairs needed during the first year, time to the first breakdown, average number of hours in use per week, cost of repairs, and so on. What is the sample space here? Conceptually we could perhaps think of the sample space as consisting of all existing items of the particular type being studied. However, it's really the quantities we are measuring rather than the physical items themselves that we are interested in. The measurements can be thought of as numerical values associated with each item. For example, the time until the first repair is needed might be 6 months for one item and 3 years for another. In any case we are associating a number with each item, and that is what a function does. We will see that by phrasing our discussions in terms of random variables, we can often bypass complicated sample spaces and instead do all our work with real numbers where everything is simpler.

The common practice in studying probability is to use uppercase letters from near the end of the alphabet such as X, Y, or Z to denote random variables. While this may seem strange at first, it does enable us to reserve common letters such as f, g, and so on, for use in the conventional calculus sense, that is, as real-valued functions of a real variable.

Example 3.1. A pair of dice is rolled. Let X denote the random variable that indicates the sum on the two dice. Using the usual 36 element sample space to represent the experiment, the value that X takes on for the outcome (2, 4) is 6. The domain of the random variable X is the 36-element sample space, and the range of X is the set of numbers $\{2, \cdots, 12\}$. One might indicate the action of X on (6, 4), for example, by writing

$$X: (6, 4) \rightarrow 10$$

(One can of course think of functions in terms of inputs and outputs. In the case of the random variable X, if we input the element (6, 4) from the sample space S, then X outputs the value 10.)

Example 3.2. A room full of people is considered as the sample space. The random variable Y gives the weight of each person; that is, when a given person is "input" to Y, the random variable Y "outputs" the weight of the person.

Example 3.3. A single die is tossed. Consider the random variable that indicates the number that appears on the die. Here it is natural to identify the outcomes of the experiment with the numbers $1, \cdots, 6$; that is, it is reasonable to consider the sample space to be the set of numbers $S = \{1, 2, 3, 4, 5, 6\}$. If one does this, then X is simply the identity function on S; that is, $X(w) = w$ for each element $w \in S$. This makes sense in this case, because S is a set of numbers.

Example 3.4. A coin is tossed twice. Let X be the number of heads that occurs. If we take $S = \{\text{HH, HT, TH, TT}\}$, then $X(\text{HH}) = 2$, $X(\text{HT}) = 1$, $X(\text{TH}) = 1$, and $X(\text{TT}) = 0$. This is similar to Example 3.1.

Someone may take the following view, however. If the number of heads is all we are interested in, why not record just that information; that is, why not consider the sample space to be just $S = \{0, 1, 2\}$, where 0 represents the outcome of no heads, 1 represents the outcome of 1 head, and 2 represents the outcome of 2 heads? In this way we can, in effect, consider the random variable that counts the number of heads to be simply the identity function $X(w) = w$ on this sample space.

This latter viewpoint is not really as different as it might first seem. We shall soon see that what is really needed in order to understand a random variable is knowledge of the *probabilities* of the random variable assuming specific values. Furthermore, the above interpretations are consistent in this respect if consistent probability measures are used on each of the two possible sample spaces. For example, the probability of the event $\{w : X(w) = 1\}$ is 1/2 in either case. In the former sample space, this event is {HT, TH}. In the latter it is $\{1\}$. The latter sample space is not a uniform one since two of the three outcomes listed have probability 1/4 and the other 1/2.

Example 3.5. Four bits of binary data are transmitted, so the sample space is that of Example 1.8. Consider Z to be the random variable that assigns to any 4-bit pattern the number of 1's appearing among the 4 bits. So

$$Z : 0110 \to 2$$

If we identify the sample space here with the numbers $0, \cdots, 15$, then the random variable is simply assigning to each number the number of 1's that appear in its binary representation.

Example 3.6. Five coins are tossed. The random variable Y indicates how many heads are obtained. Here Y acts on a uniform sample space of $32 = 2^5$ possible outcomes for the experiment and Y assumes values $\{0, 1, 2, 3, 4, 5\}$.

Example 3.7. Lines of text are being transmitted. Each line is 80 characters. We will suppose that for formatting purposes it is necessary to know how many blank spaces are included in each line. Furthermore, we assume that characters may consist of the 26 letters, 10 digits, or 6 different punctuation symbols, for a total of 42 different characters.

The number of different possible lines that could be transmitted would then be equal to 80^{42}. If we take this set of all possible lines as our sample space and if we wish to count the blank characters, then we should think of the random variable defined on this sample space that simply assigns to each possible line the number of blank characters within the line. Clearly the values that this random variable can assume are restricted to the counting numbers from 0 (no blanks in a line) to 80 (the whole line being made up of blank spaces).

Example 3.8. A resistor is put into a circuit and the length of time X it lasts before burning out is measured. This situation is a bit more vague since it is not clear what the "possible outcomes" are. As a first effort one might try to visualize some hypothetical batch of resistors as the sample space. Since it is "time to burnout" that we are interested in, however, it is perhaps more useful to think of the sample space as a set of numbers representing possibilities for the time to burnout. In fact, one could construe any non-negative length of time as being hypothetically possible, in which case the random variable could take on all non-negative values. With this point of view, the random variable X is simply the identity function, in other words $X(w) = w$ for all $w \geq 0$. Clearly in this example we have no guess as to how to treat probabilities on such a sample space. That will come later.

It is useful to have a short notation for certain events associated with random variables. For instance, in Example 3.1, the event $\{w : X(w) = 4\}$ is simply the set $\{(1, 3), (2, 2), (3, 1)\}$. The notation $\{w : X(w) = 4\}$ is often shortened to $(X = 4)$. Similarly, $\{w : X(w) < 9\}$ is shortened to $(X < 9)$. Other similar situations are handled in a like manner. And finally, the probabilities of the two events $(X = 4)$ and $(X < 9)$ are usually written as $P(X = 4)$ and $P(X < 9)$.

3.1 Discrete Random Variables and Probability Mass Functions

Random variables come in two primary "flavors," *discrete* and *continuous*. A random variable that assumes only finitely or countably infinitely many different values is a discrete random variable. (A countably infinite set is a set that can be put in one-to-one correspondence with the positive integers.) The most common discrete random variables are integer-valued and generally arise in situations where some kind of counting is being performed. The important information about a discrete random variable is carried by its *probability mass function*.

Definition 3.2: Given a discrete random variable X, the *probability mass function* p_X for X is defined for all numbers t by $p_X(t) = P(X = t)$.

Example 3.9. If X is the sum when a pair of dice is rolled, then X has range equal to the set $\{2, \cdots, 12\}$. So $p_X(t)$ takes on non-zero values only when $t \in \{2, \cdots, 12\}$. For example, $p_X(5) = 4/36 = 1/9$, whereas $p_X(13) = 0$.

Example 3.10. If X is the number of 1's that appear when 6 random binary digits are transmitted, then $p_X(t) \neq 0$ only when $t \in \{0, 1, 2, 3, 4, 5, 6\}$. If we think of the stream of digits as an independent trials process, then the values of the probability mass function are easily obtained from the independent trials formula. For example, the probability of exactly two 1's being transmitted among the 6 digits is $p_X(2) = C(6, 2)\,(1/2)^2\,(1/2)^4$.

Example 3.11. If X is the number of heads to occur in two tosses of a coin, then $p_X(0) = p_X(2) = 1/4$ and $p_X(1) = 1/2$. Notice in Example 3.4 that this is true for either proposed sample space.

One of the important attributes of the probability mass function is that it incorporates the important information about the values the random variable assumes, and the probabilities of these values, without reference to the sample space at all. Questions about the random variable can be answered by using just the mass function.

Example 3.12. If 10 random binary digits are transmitted, what is the probability that more than seven 1's are included among them?

Solution: If we let X denote the number of 1's among the 10 digits, then what we have to compute is

$$
\begin{aligned}
P(X > 7) &= P(X = 8) + P(X = 9) + P(X = 10) \\
&= p_X(8) + p_X(9) + p_X(10) \\
&= C(10, 8)\left(\frac{1}{2}\right)^8\left(\frac{1}{2}\right)^2 + C(10, 9)\left(\frac{1}{2}\right)^9\left(\frac{1}{2}\right) + C(10, 10)\left(\frac{1}{2}\right)^{10} \\
&= (45 + 10 + 1)\left(\frac{1}{2}\right)^{10} = \frac{56}{1024}
\end{aligned}
$$

3.2 Continuous Random Variables and Density Functions

One can intuitively think of a continuous random variable as one that takes on values in some interval of the real line. For example, many continuous random variables have as their range of possible values the non-negative real numbers. The concept of the probability mass function does not apply to random variables of this type. The analogous concept is that of the *density function* of the random variable. Fundamentally, when dealing with continuous random variables, one has to ask a different kind of question. Rather than asking for the probability that the random variable assumes a specific value, we ask instead what is the probability that it assumes a value in some interval of numbers. The role that the probability mass function plays in the discrete case is played by the density function in the continuous case.

Definition 3.3: A *density function* is a function f defined on the real numbers and having the following properties:

1. $f(t) \geq 0$ for all t
2. $\int_{-\infty}^{\infty} f(t)\, dt = 1$

Since the endpoints of the integration in Equation 2 are infinite, we must be prepared to deal with improper integrals. Often, however, these integrals may be reduced to integrals over an interval of finite length. For example, if a density function happens to have the property that $f(t) = 0$ when t lies outside of some interval $[a,b]$ of numbers, then

$$\int_{-\infty}^{\infty} f(t)\, dt = \int_{a}^{b} f(t)\, dt$$

The analogy between this definition and the definition of the probability mass function in the discrete case is that the requirement that the integral equals 1 in Definition 3.3 corresponds to the fact that the sum of the (nonzero) values of a probability mass function is 1. Whereas each discrete random variable has a *probability mass function*, every continuous random variables has a *density function*. The relationship between a continuous random variable and its density function is given in Definition 3.4.

Definition 3.4: A random variable X is said to have density function f provided that f is a density function related to X in the following way:

For all numbers a and b with $a < b$,

$$P(a < X < b) = \int_a^b f(t)\, dt$$

The values $a = -\infty$ and $b = \infty$ are permitted here, in which case this becomes an improper integral.

If X is a random variable with a density function, it is common to denote the density function by f_X. Sometimes when there can be no confusion from context, we will drop the subscript and just write f. For the purposes of this book, it is sufficient to think of a *continuous random variable* as simply a random variable that has a density function.

Example 3.13. Perhaps the most intuitively natural density function is the function f defined by $f(t) = 1$ for $0 < t < 1$ and $f(t) = 0$ otherwise. If a and b are two numbers with $0 \le a < b \le 1$, then

$$P(a < X < b) = \int_{-\infty}^{\infty} f(t)\ dt = \int_a^b 1\ dt = b - a$$

Thus the probability of X assuming a value in a given subinterval of the interval [0, 1] is simply the length of that subinterval.

This provides a continuous model for the experiment of "choosing a random number between 0 and 1."

Example 3.14. Figure 3.1 shows the density functions for three random variables. Suppose X_1 has f_1 as its density function and X_2 has f_2. Since both functions vanish off the interval [0,1], we know that $P(X_1 < 0) = 0$ and also that $P(X_1 > 1) = 0$ and similarly for X_2. In other words, X_1 and X_2 are random variables that assume values between 0 and 1. The graph of f_1, however, is higher toward the right end of this interval, and the graph of f_2 is higher toward the left end. This tells us that values for X_1 are more probable near 1 than near 0, and vice versa for X_2. As a sample computation to illustrate this, let's compute the probabilities $P(.8 < X_1 < 1)$ and $P(0 < X_1 < .2)$.

$$P(0 < X_1 < .2) = \int_0^{.2} 2t\,dt = .04$$

whereas,

$$P(.8 < X_1 < 1) = \int_{.8}^{1} 2t\,dt = .36$$

The situation regarding X_2 is exactly reversed.

Example 3.14 illustrates that a simple glance at the density function gives much information about the random variable. A random variable having the function g at the bottom of Figure 3.1 as its density function would be one that assumes values between 2 and 5 with values around 3 being most probable.

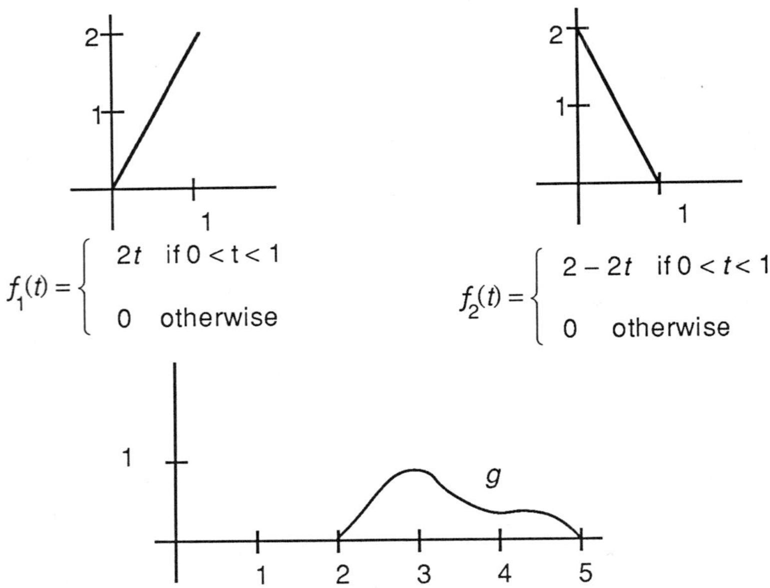

Figure 3.1 Three density functions.

An important observation is that *any random variable X having a density function automatically has the property that for every number t_0, $P(X = t_0) = 0$.*

To see this, simply think of taking a small interval of radius δ centered at t_0. Then

$$P(X = t_0) \leq P(t_0 - \delta < X < t_0 + \delta) = \int_{t_0 - \delta}^{t_0 + \delta} f_X(t)\, dt$$

and the fact that f is an integrable function guarantees that the integral on the right tends to zero as $\delta \to 0$. Since $P(X = t_0)$ is less than or equal to this quantity for every positive δ, $P(X = t_0)$ must equal zero.

Proposition 3.1 includes the observation we have just made as well as a useful additional consequence of this fact.

Proposition 3.1: If X is a random variable having a density function, then for any number t_0, $P(X = t_0) = 0$, and for any numbers a and b,

$$P(a < X < b) = P(a \leq X < b) = P(a < X \leq b) = P(a \leq X \leq b)$$

The explanation for the latter part of Proposition 3.1 is simple. For example,

$$\begin{aligned} P(a \leq X < b) &= P(a < X < b) + P(X = a) \\ &= P(a < X < b) + 0 = P(a < X < b) \end{aligned}$$

The other parts may be proved in a similar way.

Other examples of commonly occurring density functions will be given in Chapter 5. Before looking at continuous models, however, it is a good idea to get a better understanding of the mathematical implications of the relation between a random variable and its density function. It is important to understand some parallels between discrete and continuous random variables, and one way to understand the parallelism is in terms of a third way of representing the "probability distribution" of a random variable: the *cumulative distribution function.*

3.3 Distribution Functions

Definition 3.5: For any random variable X (discrete, continuous, or neither), the *cumulative distribution function* of X (or simply the *distribution function* of X) is the function F_X defined on the real numbers by

$$F_X(t) = P(X \leq t) \quad \text{for all } t$$

It is useful to understand what the distribution function for a discrete random variable looks like and how it relates to the probability mass function. Figure 3.2 shows the graph of the distribution function for the random variable X that counts the number of heads in three coin tosses. This figure illustrates the fact that for a discrete random variable, the distribution function is a step function with a jump discontinuity at each value of t such that $p_X(t) \neq 0$. Furthermore, the magnitude of the jump discontinuity is equal to $p_X(t)$.

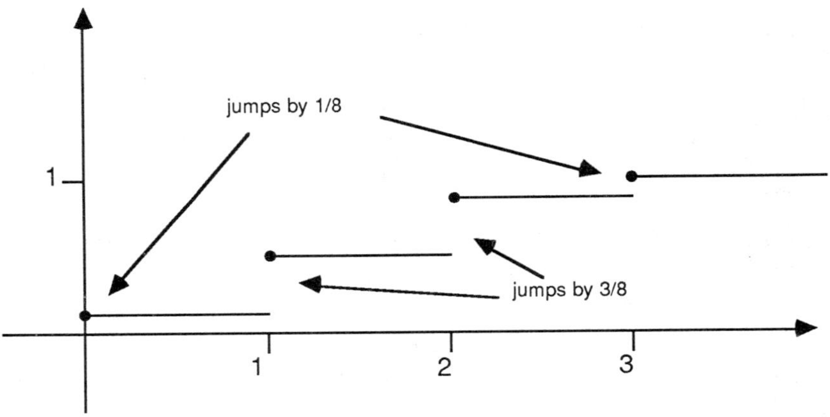

Figure 3.2 Distribution function for the random variable that counts the number of heads in 3 coin tosses.

Proposition 3.2: If a function F is the distribution function for some random variable, then F has the following properties:

1. $F(t) \to 1$ as $t \to \infty$.
2. $F(t) \to 0$ as $t \to -\infty$.
3. F is right-continuous on the real line; that is,

$$\lim_{t \to t_0+} F(t) = F(t_0) \quad \text{for every } t_0$$

4. F is a nondecreasing function; that is, if $t_1 < t_2$ then $F(t_1) \leq F(t_2)$.

A proof of Proposition 3.2 may be based on the basic properties of probabilities found in Chapter 1. The first three parts of Proposition 3.2 are limit properties and require use of Axiom 3.1 in Chapter 1.

One should be very aware that the word *distribution* is often used in a very loose sense. In light of Proposition 3.3, it is straightforward to pass from the density function to the distribution function, and vice versa. The term *distribution* is sometimes used to describe a density function and sometimes a distribution function. For example, one might say that Definition 4.1 in the next chapter describes the "binomial distribution," although literally it is the probability mass function of a binomial random variable that is being described. Similarly, it might be said that Definition 5.2 encountered in Chapter 5 describes the "exponential distribution," though in fact it is the density function that is defined there.

In the case of a continuous random variable, the fundamental theorem of calculus gives the relation between the density function and the distribution function. The relation is shown in Proposition 3.3. The important point here is that the distribution function and the density function for a continuous random variable may easily be derived from each other. If either one is known, the computation of the other boils down to an integration or a differentiation. This frequently makes calculations with continuous random variables easier than calculations with discrete ones.

Proposition 3.3: Suppose X is a random variable having density function f_X and distribution function F_X. Then these two functions are related as follows:

1. $F_X(t_0) = \int_{-\infty}^{t_0} f_X(t)\, dt$

2. $\frac{d}{dt}[F_X(t)] = f_X(t)$ for all values of t where f_X is continuous

---

The similarity between the discrete case and the continuous case is illustrated by Figure 3.3. The integrals encountered when working with continuous random variables replace the sums that appear in the discrete case, and the density function replaces the probability mass function.

Expression	Discrete case	Continuous case
$P(a \le X < b)$	$\sum_{a \le t_k < b} p_X(t_k)$	$\int_a^b f_X(t)\ dt$
$F_X(t_0)$	$\sum_{t_k \le t_0} p_X(t_k)$	$\int_{-\infty}^{t_0} f_X(t)\ dt$

Figure 3.3 Integration replaces summation in the case of continuous random variables.

Example 3.15. Suppose X is a random variable having density function defined by $f(t) = 2t$ for $0 < t < 1$, with $f(t) = 0$ otherwise. (This is the function f_1 of Figure 3.1.) Find the distribution function F_X.

Solution:

$$F_X(s) = \int_{-\infty}^{s} f(t)\ dt = 0 \text{ if } s < 0$$

$$F_X(s) = \int_0^s 2t\ dt = s^2 \text{ if } 0 \le s \le 1$$

$$F_X(s) = \int_0^1 2t\ dt = 1 \text{ if } s \ge 1$$

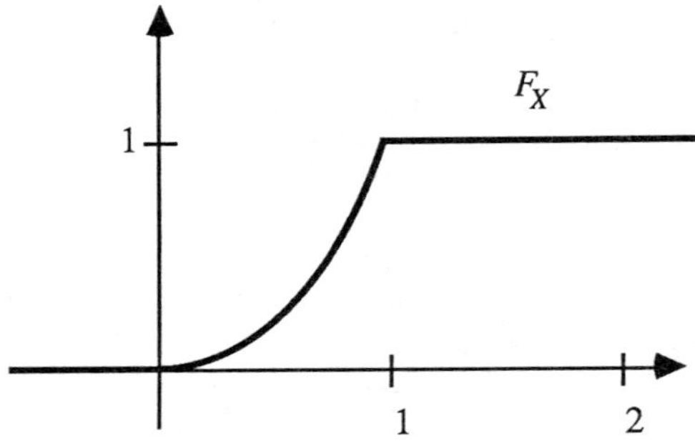

Notice that the derivative of F_X is f_X at all points where f_X is continuous, which in this example is everywhere except at $t = 1$. Furthermore, notice that F_X is a continuous function, whereas f_X is not (because of its discontinuity at $t = 1$). The distribution function for a continuous random variable will *always* be a continuous function. The density function may or may not be.

A little should be said about the "not quite uniqueness" of the density function for a random variable. Since the relation between a random variable and its density function depends solely on integrating the density function, strictly speaking the density function is not unique. (Two different functions can be density functions for the same random variable.) To see this, simply think about changing the value of the density function at some finite set of numbers. Making such a change would not change the value of the integral of the function over any interval. For example, in the case of the uniform density on the interval [0,1], it doesn't matter whether we say $f(t) = 1$ on the closed interval $0 \leq t \leq 1$ or the open interval $0 < t < 1$. The only difference in the two involves the value of $f(t)$ at the endpoints 0 and 1, and the value assigned at these two places will not affect the value of the integral of f over any subinterval of the real line. These observations notwithstanding, it is a common practice to speak of *the* density function of a random variable with the realization that this is a slight abuse of language.

When dealing with continuous random variables, Proposition 3.1 says that one can disregard the endpoints of intervals since the probability of the random variable assuming a value in a given interval is the same regardless of whether the endpoints are included or not. With discrete random variables this is not the case. For a discrete random variable X, $P(a < X < b)$ and $P(a \leq X \leq b)$ are not the same in general, the reason being simply that there is no longer any guarantee that $P(X = a)$ and $P(X = b)$ are 0. In fact, it is easy to see that $P(a < X < b) = P(a \leq X \leq b)$ if and only if $P(X = a) = P(X = b) = 0$.

3.4 Expected Value and Variance

The *expected value*, or *expectation* or *mean*, of a random variable is a weighted average of the values assumed by the random variable. The notation commonly used for the expected value of a random variable X is $E(X)$. For example, if X is the number obtained when a fair die is rolled, then the expected value is $E(X) = 3.5$, which is just the average of the numbers 1 to 6 that constitute the values of X.

If Y is the number of defects in a sample of two devices drawn from a batch of 5 of which 2 are defective, then it is easy to check (using a tree diagram) that $P(Y = 0) = .3$, $P(Y = 1) = .6$, and $P(Y = 2) = .1$. In this case $E(Y)$ is defined as

$$E(Y) = 0 \times .3 + 1 \times .6 + 2 \times .1 = .8$$

This example illustrates that the expected value of a random variable is indeed a weighted average, where the weights are simply the probabilities of the respective outcomes. The rationale is that if an experiment were conducted a large number of times, each leading to an "observed value" of the random variable and if each value assumed by the random variable were to be "observed" exactly as frequently as predicted by the probability mass function, then this theoretical weighted average would indeed match the true average of all the observed results of the sequence of experiments. The fact that $E(Y) = .8$ in this simple illustration reflects the fact that if the experiment were repeated a large number of times and if no defects were found 30% of the time, 1 defect was found 60% of the time, and 2 were found 10% of the time, then the average number found would be .8. This idea is incorporated in Definition 3.6.

Definition 3.6: If X is a discrete random variable, then the *expected value*, or *mean*, of X is defined by

$$E(X) = x_1 P(X = x_1) + x_2 P(X = x_2) + \cdots$$

where $x_1, x_2, \cdots$ represent the values X assumes.

In case X assumes infinitely many values, then the sum on the right is to be interpreted as the sum of an infinite series.

If X assumes infinitely many values, then there is no guarantee that the series in the above definition converges. If the series fails to converge, then the mean of the random variable is not defined. As a matter of fact, when one speaks of a random variable having an expected value, it is generally understood that not only does the series above converge, but that it converges absolutely.

If we keep in mind that for continuous random variables integration replaces summation, the following definition should be expected (no pun intended).

Definition 3.7: If X is a continuous random variable with density function f, then the *expected value* of X is defined by

$$E(X) = \int_{-\infty}^{\infty} tf(t)\, dt$$

Just as with the sum in the discrete case, there is no guarantee that the improper integral here will converge. Thus when one says that a given continuous random variable has a mean or expected value, the statement implies that the integral converges, and again common usage is to interpret this as meaning that the integral actually converges absolutely.

Example 3.16. Suppose X is a random variable having as its density function the function f_1 of Figure 3.1. The expected value of X then is given by

$$E(X) = \int_{-\infty}^{\infty} tf_1(t)\, dt = \int_0^1 t\,(2t)\, dt = 2 \times \frac{t^3}{3}\Big]_0^1 = \frac{2}{3}$$

A very important property of expected value is the property of being a linear operation. Proposition 3.4 states this linearity property.

Proposition 3.4: If X and Y are random variables (with finite expected value) defined on the same sample space and if a and b are real numbers, then

$$E(aX + bY) = a\,E(X) + b\,E(Y)$$

The proposition asserts that the expected value of $aX + bY$ will exist automatically (the sum or integral will converge) if X and Y have expected values. In this case, the proposition says that the expected value of a linear combination of X and Y is the corresponding linear combination of their expected values.

This fact is actually deeper than it might appear. For a finite or countably infinite sample space, the proposition can be proved by first showing that

$$E(Z) = \sum Z(w_k)\, P(\{w_k\})$$

for any random variable Z, where the sum extends over all elements w_k in the sample space. For then,

$$\begin{aligned} E(aX + bY) &= \sum (aX + bY)(w_k)\, P(\{w_k\}) \\ &= a \sum X(w_k)\, P(\{w_k\}) + b \sum Y(w_k)\, P(\{w_k\}) \\ &= a\, E(X) + b\, E(Y) \end{aligned}$$

For continuous random variables there is no elementary proof of Proposition 3.4. In higher level treatments of probability theory (treatments preceded by a study of real variables, including measure theory), Proposition 3.4 arises naturally in the theory of integration. In any case, it should be apparent that Proposition 3.4 can be used repeatedly to apply to any linear combination of a finite number of random variables.

Example 3.17. A construction project requires 30 components. Of these, 10 have been acquired from manufacturer A and 20 from manufacturer B. If 8 of the components are randomly chosen to be used in the first stage of construction, what is the expected value for the number of these that come from manufacturer A?

Solution: If we consider X to be the number among these 8 that come from manufacturer A, it is not terribly difficult to determine the probability mass function for X. Clearly X can assume values $0, 1, \cdots, 8$, and the probability of each of these values can be computed using Proposition 1.2, the combinations formula. The number of different ways that 8 components can be chosen from the 30 is $C(30, 8)$. To compute the probability that, for example, 3 of these come from manufacturer A, we need to know how many of these $C(30, 8)$ ways of choosing the components for stage one would involve using 3 of manufacturer A's components and 5 of manufacturer B's. The number of ways to pick 3 of A's is $C(10, 3)$, and the number of ways to pick 5 of B's is $C(20, 5)$. Thus the number of ways to use 3 of A's and 5 of B's is $C(10, 3) \times C(20, 5)$. Dividing this number by the total number of possible outcomes, $C(30, 8)$, would give $P(X=3)$. If this method is used to compute $P(X = 0), \cdots, P(X = 8)$, then

$E(X)$ may be determined from Definition 3.6.

Proposition 3.4, however, offers a simple alternative. Let's think of the 8 components to be used in stage one as being chosen one at a time. (This is a mental trick; it makes no difference whether they are actually chosen one at a time or simultaneously. Is that clear to you?) We will consider X_1 to be 1 if the first one chosen comes from manufacturer A and 0 if it comes from manufacturer B. Similarly $X_2, \cdots, X_8$ tell whether the others among the 8 being used come from manufacturer A. So $X_1, X_2, \cdots, X_8$ are very simple random variables. Each has probability 1/3 of assuming value 1 and 2/3 of assuming value 0 since 1/3 of the total come from A and 2/3 from B. So $E(X_1) = E(X_2) = \ldots = 1/3$. Furthermore, the relation they have to X is that

$$X = X_1 + X_2 + \cdots + X_8$$

In light of the linearity property of expected value,

$$E(X) = E(X_1) + \cdots + E(X_8)$$

Each of the expected values on the right is 1/3, and so this means $E(X) = 8/3$. A little reflection will tell you that this is the only reasonable answer. For if 1/3 of the total available components are from manufacturer A, then surely the expected number in our chosen lot of 8 should be 1/3 of these, which is 8/3.

A trick encountered in this example is so useful that it deserves special comment. Often it is helpful to have a random variable that serves as a flag to indicate whether a particular event occurs or not. The procedure used in the last example was to introduce a random variable that takes on value 1 if the event occurs and value 0 if it does not. While other values could be used besides 0 and 1, the nice thing about that choice in the preceding example was that to find how many of the events occurred we then needed only to add the values of the corresponding random variables. Given an event E, it is common practice to refer to the random variable that takes on value 1 if E occurs and value 0 if E does not occur as the *indicator random variable* for the event E. We will have occasion to use this trick from time to time in future chapters.

For a given random variable X, $E(X)$ is a number that gives a useful characteristic of the random variable. An alternate notation that is frequently used for $E(X)$ is μ_X. In the study of statistics, μ_X is often referred to as the *mean* of X. Another useful number is the *variance* of the random variable.

Definition 3.8: If X is a random variable with expected value μ_X, then the *variance* of X is defined by

$$\text{var}(X) = E[\,(X - \mu_X)^2\,]$$

The *standard deviation* of X is the square root of var(X). It is common practice to denote the standard deviation of X by σ_X and the variance by σ_X^2.

Whereas the mean represents the theoretical average value, the *variance* indicates the extent to which values of the random variable tend to concentrate around the average (leading to a small variance) or fluctuate greatly (leading to a large variance). The variance or standard deviation is used as a measure of how "spread out" the values of a random variable are. If the values of X tend to cluster tightly around the mean μ_X, then the variance of X will be small. On the other hand, if the values of X vary widely with high probability, then the variance will be large. It is of course possible that $(X - \mu_X)^2$ not have finite expected value, which is just to say that not all random variables have a finite variance.

Definition 3.8 is useful in that it gives a good intuitive understanding of what it is that the variance measures. There is an alternate way of computing the variance, however, that is often more convenient. This alternative is given in Proposition 3.5. Proposition 3.5 says that if we know the mean, then all we need to know additionally is $E(X^2)$.

Proposition 3.5: The variance of a random variable X is given by

$$\text{var}(X) = E(X^2) - [E(X)]^2 \quad \text{and} \quad \text{var}(aX) = a^2\,\text{var}(X)$$

Proof:

$$\begin{aligned} E[(X - \mu_X)^2] &= E(X^2 - 2\mu_X X + \mu_X{}^2) \\ &= E(X^2) - 2\mu_X E(X) + \mu_X{}^2 \\ &= E(X^2) - 2\mu_X{}^2 + \mu_X{}^2 \\ &= E(X^2) - [E(X)]^2 \end{aligned}$$

Also,

$$\begin{aligned} \text{var}(aX) &= E[(aX)^2] - [E(aX)]^2 \\ &= E(a^2X^2) - [aE(X)]^2 \end{aligned}$$

$$= a^2E(X^2) - a^2E(X)^2$$
$$= a^2 \text{ var}(X)$$

While most random variables encountered in practice fall into one of the categories *discrete* or *continuous*, it is not hard to visualize situations where a mixture of the two might occur. First, let's examine a situation that leads to a mix of two continuous random variables of the type found in Problem 3.14.

Example 3.18. Two different manufacturers supply a component with an exponentially distributed lifetime; that is, the length of service the component gives until it fails is an exponentially distributed random variable with a density function of the type found in Problem 3.14. Manufacturer A's device has expected lifetime 4 months and manufacturer B's has 10 months. A particular user has a batch of devices of which 40% came from manufacturer A and 60% from manufacturer B. If a randomly selected device from this batch is used, what is the probability distribution for its lifetime and what is the expected value for its lifetime?

Solution: We will denote by X the random variable that gives the time to failure for the randomly selected device. What we must do is to partition the sample space according to the manufacturer of the chosen device. If we think of E and F as the following two events,

E = event that device is made by manufacturer A

F = event that device is made by manufacturer B

then for any $t > 0$,

$$P(X > t) = P(E)\,P(X > t \mid E) + P(F)\,P(X > t \mid F)$$
$$= .4\,e^{-.25t} + .6\,e^{-.1t}$$

The exponential terms in this expression come from the simple calculation called for in Problems 3.14 and 3.15. (By Problem 3.15, the expected value of an exponential random variable is the reciprocal of the parameter λ in the distribution. The reciprocal of 4 is .25 and the reciprocal of 10 is .1). Thus the distribution function for X is given by

$$F_X(t) = 1 - .4\,e^{-.25t} - .6\,e^{-.1t} \quad \text{when } t > 0$$

The density function then is

$$f_X(t) = .1\,e^{-.25t} + .06\,e^{-.1t} \quad \text{for } t > 0$$

Problem 3.20 asks you to find the expected value for the random variable X, that is, the expected time to failure for the randomly chosen component under discussion.

Example 3.19. We will now consider a variation on the problem just considered. Suppose that manufacturer B leases its components rather than selling them. What one buys then is a 10-month service contract rather than a single component. If the original fails during that period, the manufacturer replaces it at no additional charge, but whatever unit is in service is removed at the end of 10 months. In this case, we know exactly how much service will result from a purchase from manufacturer B.

If we continue to assume that 40% of a given user's purchases are components from manufacturer A and 60% are 10-month service contracts from manufacturer B, what now is the probability distribution for the amount of service obtained from a randomly chosen purchase?

Solution: Again we can let E and F be described by the following:

E = event that the purchase is a component from manufacturer A

F = event that the purchase is a service contract from manufacturer B

Then the length of time of service provided can be thought of as a random variable X, and for any number $t > 0$,

$$P(X > t) = P(E)\,P(X > t \mid E) + P(F)\,P(X > t \mid F)$$

The first term on the right is still $.4\,e^{-.25t}$ for the same reasons as in the previous example. However, the term $P(X > t \mid F)$ will be 1 if $t < 10$ and will be 0 if $t \geq 10$. So we have

$$P(X > t) = \begin{cases} .4\,e^{-.25t} + .6 & \text{if } t < 10 \\ .4\,e^{-.25t} & \text{if } t \geq 10 \end{cases}$$

The distribution function for X is then given by

$$F_X(t) = 1 - P(X > t) = \begin{cases} .4\,(1 - e^{-.25t}) & \text{if } t < 10 \\ 1 - .4\,e^{-.25t} & \text{if } t \geq 10 \end{cases}$$

The graph of this distribution function is shown in the following picture. From the form of the equations defining $F_X(t)$, it is apparent that there is going to be a jump discontinuity of magnitude .6 at $t = 10$.

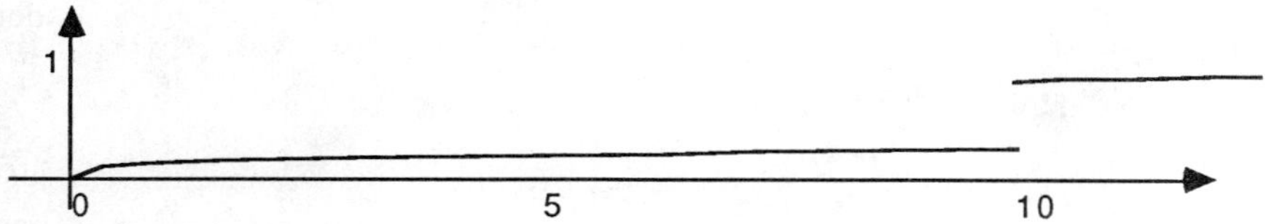

If you bear in mind that a discrete random variable always has a step function as its distribution function and that a continuous random variable always has a continuous function as its distribution function, it is apparent here that we are looking at a distribution function that is of neither type. In particular, there will be no probability mass function or density function for this random variable. For distributions of this "mixed" type, the cumulative distribution function is the only tool available in computations.

3.5 Functions of a Random Variable

A random variable is simply a real-valued function defined on some sample space S. This means that if g is some ordinary real-valued function defined on some part of the real line that includes the range of the random variable X, we can consider the random variable $g(X)$ which is the composite of the two functions.

Example 3.20. If X is the number obtained when a die is rolled and g is the function $g(t) = t^2$, then $g(X) = X^2$ is the random variable (defined on the same sample space as X) that gives the square of the number obtained when the die is rolled. The values that $g(X)$ assumes are 1, 4, 9, 16, 25, and 36, each with probability 1/6.

Example 3.21. If X is the random variable that represents a random number drawn from the unit interval [0, 1], and $g(t) = t^2$, then $g(X) = X^2$ is the random variable that would be used to model "squaring a random number between 0 and 1."

Example 3.22. Suppose $S = \{w_1, w_2, w_3\}$ and $X(w_1) = 3$, $X(w_2) = 5$, and $X(w_3) = -3$. Furthermore let's assume that S is a uniform sample space in which each of the elements of S has probability 1/3.

Now consider the composite random variable $Y = g(X)$, where $g(t) = t^2$, that is, $Y = X^2$. It is easy to see that Y assumes only two values, 9 and 25. Thus, according to the definition of expected value,

$$E(Y) = 9\,P(Y = 9) + 25\,P(Y = 25) = 9 \times \frac{2}{3} + 25 \times \frac{1}{3} = \frac{43}{3}$$

This computation involves the probability mass function of Y. It is very useful to notice that $E(Y)$ can be computed by using the mass function of X. For in fact

$$E(Y) = 9\,P(X = 3) + 9\,P(X = -3) + 25\,P(X = 5)$$

Recall that the definition of $E(Y)$ requires us to know the probability mass function for Y itself. The last expression for $E(Y)$ given here depends very heavily on the fact that Y is a function of X. This method, as shown in Proposition 3.6, can be applied when one random variable is expressed as a function of another.

Proposition 3.6: If X is a discrete random variable and if $Y = g(X)$, then

$$E(Y) = \sum g(x_k)\,P(X = x_k)$$

where the sum extends over all values x_k assumed by the random variable X.

The importance of this proposition is in situations where the probability mass function of X is well understood, and so the sum appearing in Proposition 3.6 can be computed to obtain $E(Y)$. If Definition 3.6 were used instead to compute $E(Y)$, it would first be necessary to know the probability mass function for Y itself. This can all be summed up by saying the following: *If Y is a function of X, then $E(Y)$ can be computed using p_X without the necessity of knowing p_Y.*

As shown in the discussion preceding Proposition 3.6, the proof of the proposition depends only upon rearranging and combining terms in the sum which defines $E(Y)$. Each term of the form $y_j P(Y = y_j)$ in the sum that defines $E(Y)$ can be split into a sum of terms of the form $g(x_k)\,P(X{=}x_k)$ where $g(x_k) = y_j$.

In the continuous case there is a similar fact that is stated in Proposition 3.7. The proof depends on a fairly sophisticated use of knowledge from advanced calculus or real variables.

Proposition 3.7: If X is a continuous random variable and if $Y = g(X)$, then the expected value of Y is given by

$$E(Y) = \int_{-\infty}^{\infty} g(t)\, f_X(t)\, dt$$

Again, as in the discrete case, the importance of this proposition is that it allows computation of $E(Y)$ without the necessity of knowing the probability density for Y itself.

Example 3.23. Let's consider taking the square root of a random number between 0 and 1. Denote by X the random variable representing the random number between 0 and 1, and the assumption is that the probability distribution of X is the uniform distribution on [0, 1] as in Example 3.13. Let's denote $Y = \sqrt{X}$, in which case $Y = g(X)$, where g is the function $g(t) = \sqrt{t}$.

According to Proposition 3.7,

$$E(Y) = \int_{-\infty}^{\infty} \sqrt{t}\, f_X(t)\, dt = \int_0^1 \sqrt{t}\, dt = \frac{2}{3}$$

In this example it is not difficult to work out the density function for Y and find the value of $E(Y)$ from Definition 3.7. In fact, for any $t \geq 0$,

$$F_Y(t) = P(Y \leq t) = P(\sqrt{X} \leq t) = P(X \leq t^2) = F_X(t^2)$$

In Problem 3.3 the distribution F_X is determined. From that exercise it follows that

$$F_Y(t) = \begin{cases} 0 & \text{if } t \leq 0 \\ t^2 & \text{if } 0 \leq t \leq 1 \\ 1 & \text{if } t \geq 1 \end{cases}$$

Differentiation of F_Y gives the density function for Y:

$$f_Y(t) = \begin{cases} 2t & \text{if } 0 < t < 1 \\ 0 & \text{otherwise} \end{cases}$$

Now $E(Y)$ can be computed from Definition 3.7:

$$E(Y) = \int_{-\infty}^{\infty} t f_Y(t)\, dt = \int_0^1 t(2t)\, dt = \frac{2}{3}$$

Except in elementary cases, such as the above, it is often not easy to determine the density function for $Y = g(X)$ even if g is simple and X has an easily described density function. Proposition 3.7 is very valuable in enabling computation of $E(Y)$ without needing to know the density function for Y.

Example 3.24. Suppose X is a random variable having density function equal to the unbounded density function

$$f(t) = \begin{cases} \dfrac{1}{2\sqrt{t}} & \text{if } 0 < t < 1 \\ 0 & \text{otherwise} \end{cases}$$

of Problem 3.19(a). What is the variance of X?

Solution:

$$E(X) = \int_0^1 t \frac{1}{2\sqrt{t}}\, dt = \frac{1}{3} \quad \text{and} \quad E(X^2) = \int_0^1 t^2 \frac{1}{2\sqrt{t}}\, dt = \frac{1}{5}$$

Therefore $\text{var}(X) = 1/5 - 1/9 = 4/45$.

Problems

3.1 Graph the cumulative distribution function for the random variable that counts the number of heads in four coin tosses.

3.2 Often it is necessary to construct a probability distribution based on a set of data. For example, suppose that 100 people have been checked by a dentist, and the breakdown on the number of cavities found is as follows.

No. of cavities	No. of people with this many cavities
0	37
1	22
2	17
3	15
4	5
5	2
6	0
7	2

Let's consider the experiment described as "randomly choosing a person from this group and counting the number of cavities." If X denotes this random variable, then clearly the probability mass function for X should be $P(X = 0) = .37$, $P(X = 1) = .22$, and so on. Sketch a graph of the distribution function for this random variable.

3.3 Suppose X is a random variable with density function equal to the uniform density function on the interval $[0, 1]$ as in Example 3.13. Determine the distribution function of X and sketch its graph.

3.4 A continuous random variable X has density function defined by $f(t) = 1/2$ for $-1 < t < 1$, with $f(t) = 0$ for other values of t.
(a) Compute $P(0 < X < .5)$.
(b) Find the value $F_X(.5)$.
(c) Now find a formula for $F_X(t)$ for all t with $-1 < t < 1$.
(d) Sketch a graph of the distribution function of X.

3.5 Determine the probability mass function and sketch a graph of the distribution function for the random variable that gives the largest number appearing when a pair of dice is rolled.

3.6 Sketch a graph of the distribution function F_X for a random variable X whose probability mass function is given by $p_X(x) = .05$ if $x = 0, .1, .2, .3, \cdots 1.8, 1.9$ and $p_X(x) = 0$ for other values of x.

3.7 If X has the density function $f(t) = 3e^{-3t}$ for $t \geq 0$, with $f(t) = 0$ whenever $t < 0$, find $P(X > 1)$ and $P(1 < X < 2)$.

3.8 Show that the function defined by $f(t) = (1/2)e^{-|t|}$ for all t is a density function, and for a random variable X having this density function, find $P(|X| < 1)$.

3.9 Suppose $f(t) = c(4 - t^2)$ for $-2 < t < 2$, with $f(t) = 0$ otherwise. Determine the value that c must have in order for f to be a density function.

3.10 Sketch a graph of the distribution function of the random variable of Example 3.9.

3.11 A word is chosen at random from the following list of words: *horse*, *dog*, *cow*, *elephant*, *pig*. ("At random" in this context means that each word has an equal chance of being chosen.) Let X be the random variable that tells the number of vowels in the word. Determine the probability mass function for X, sketch the graph of F_X, and find $E(X)$.

3.12 A factory produces transistors. Of those produced, 10% are defective. Four transistors produced by the factory are tested, and X is the random variable that tells how many defects are found. Find the probability mass function for X, and graph the distribution function. (Consider this an independent trials process in which each transistor tested constitutes a trial.)

3.13 Suppose X is a continuous random variable having density function $f(t) = 2e^{-2t}$ for $t \geq 0$, with $f(t) = 0$ for $t < 0$. Let $Y = 4X$.
(a) Compute $E(X)$ from Definition 3.7.
(b) Based on Proposition 3.4, what is $E(Y)$?
(c) Use Proposition 3.7 to compute $E(Y)$ using the density for X.
(d) Determine the density function for Y and then compute $E(Y)$ directly from Definition 3.7.

3.14 Show that if X is a random variable having the density function

$$f(t) = \begin{cases} \lambda e^{-\lambda t} & \text{for } t \geq 0 \\ 0 & \text{for } t < 0 \end{cases}$$

where $\lambda > 0$, then $P(X > t) = e^{-\lambda t}$.

3.15 Show that the expected value of a random variable having the density function of Problem 3.14 is $E(X) = 1/\lambda$.

3.16 Suppose X is uniform on $[0, 1]$. Determine and sketch a graph of the density function and the distribution function of $Y = 2X + 3$.

3.17 Suppose X is a continuous random variable and that $Y = aX + b$, where a and b are real numbers with $a > 0$. Show how to write the density function for Y in terms of the density function for X. [Hint: First determine the relation between the distribution functions, and then differentiate.] You will use this technique in exercises in Chapter 5.

3.18 A device is tested. If it is found to be defective, then two more devices from the same lot are tested. However, if the first device is found to be okay, then only one additional item is tested. Assume that the devices are independent and that they come from a batch in which 10% are defective. Let X denote the random variable that gives the total number of defective devices encountered during this testing procedure.

(a) Draw a tree diagram to represent this experiment.
(b) Find the probability mass function for X.
(c) Find the expected value of X.

3.19 Verify that each of the following is a density function:

$$\text{(a)}\ f(t) = \begin{cases} \dfrac{1}{2\sqrt{t}} & \text{if } 0 < t < 1 \\ 0 & \text{otherwise} \end{cases} \qquad \text{(b)}\ f(t) = \begin{cases} |t| & \text{if } |t| < 1 \\ 0 & \text{otherwise} \end{cases}$$

(Notice that the first of these two functions is unbounded, so the integral in Definition 3.3 must be interpreted as an improper integral.)

3.20 Verify that the function defined by $f(t) = .1e^{-.25t} + .06e^{-.1t}$ for all numbers $t > 0$, with $f(t) = 0$ for $t < 0$, is a density function and find the expected value of a random variable having such a density function. (This is the final step in Example 3.18).

3.21 Find var(X), where X is the number of heads in 4 coin tosses.

(a) Use Definition 3.8 and Proposition 3.6.
(b) Now do it using Proposition 3.5.

3.22 If X is a random variable whose density function is given by ~~$f(t) = 1/4$ whenever $-1 < t < 1$ and $f(t) = 3/4$ when $1 < t < 2$~~, sketch a graph of the cumulative distribution function for X.

$$f(t) = \begin{cases} 1/4 & -1 < t < 0 \\ 3/4 & 1 < t < 2 \end{cases}$$

Chapter 4: Discrete Models

This chapter will demonstrate how some of the most common discrete probability distributions are used to model real-world phenomena. Two of these have been presented in Chapter 1, though not by name. They are the binomial distribution and the geometric distribution. Both are intimately involved with independent trials processes.

4.1 The Binomial Distribution

Definition 4.1: A random variable X is said to be *binomial* with parameters n and p (where n is a positive integer and $0 < p < 1$) provided that X assumes values $0, 1, 2, \cdots, n$, and

$$P(X = k) = C(n, k)\, p^k\, (1 - p)^{n-k} \text{ for } k = 0, 1, 2, \cdots, n$$

Clearly the natural way in which a binomial random variable arises is in counting the number of "successes" in an independent trials process, since the expression for $P(X = k)$ is simply the independent trials formula of Chapter 1. For example, if X is the number of days in a year in which a certain electrical grid experiences a power outage, and if the probability of an outage occurring on any given day is .01, then assuming this to be an independent trials process means that X is binomial with parameters $n = 365$ and $p = .01$.

Frequently an independent trials process is called a *Bernoulli process* or a sequence of *Bernoulli trials*. (Jakob Bernoulli was a seventeenth century mathematician who wrote one of the earliest treatises on probability.) The special case of a binomial random variable for which the parameter n is equal to 1 is called a *Bernoulli random variable*.

The mean and the variance of a binomial random variable are given in Proposition 4.1. This information will be used in many applications.

Proposition 4.1: Suppose X is a binomial random variable with parameters n and p. Then

1. $E(X) = np$ and
2. $\text{var}(X) = npq$ where $q = 1 - p$

The first part of this proposition is extremely plausible. If 10 coins are tossed, the expected number of heads is 5. (Here $n = 10$ and $p = 1/2$.)

Some texts provide proofs for this proposition that depend on rather tedious observations involving binomial coefficients. It is more instructive to focus on the relationship between the first equation $E(X) = np$ and the linearity of expected value, Proposition 3.4.

Consider n trials in an independent trials process with success probability p. It is useful to introduce random variables $X_1, \cdots, X_n$ to indicate whether success occurs on each of the n trials. Specifically, let X_1 be the random variable that assigns value 1 to any outcome of the sequence of trials in which success occurs on the first trial and assigns value 0 to any outcome in which failure occurs on the first trial. Similarly, $X_2 = 1$ if success occurs on the second trial and $X_2 = 0$ if failure occurs on the second trial, etc.

Now if we look at $X = X_1 + \cdots + X_n$, the random variable X will tell how many successes occur in the n trials. If you are at all confused about what is going on here, it might be a good idea to consider a special case, such as $n = 3$. In that case the sample space can be represented by a tree diagram or can be viewed as {SSS, SSF, SFS, SFF, FSS, FSF, FFS, FFF} where S is success and F is failure. Give the value of X_1, X_2, and X_3 for each of the eight elements of the sample space, and make sure you understand that $X = X_1 + X_2 + X_3$ is indeed the number of successes in the three trials.

By the linearity property of expected value (Proposition 3.4), the expected value of X is $E(X) = E(X_1) + \cdots + E(X_n)$. However, for each k from 1 to n, X_k assumes only the values 0 and 1, and $E(X_k) = P(X_k = 1) = p$. Thus $E(X)$, being a sum of n terms each of which is equal to p, is in fact np.

A short proof of the second part of Proposition 4.1 will be given in Chapter 6 as an application of the fact that variance is additive for "independent" random variables. (The concept of independence of random variables is introduced in Chapter 6.)

4.2 The Geometric Distribution

The geometric distribution is rooted in the concept of geometric series introduced in Chapter 1. Like binomial random variables, random variables having such a distribution arise in observing independent trials processes.

Definition 4.2: A random variable X that assumes positive integer values is said to be *geometric* with parameter p provided that for each positive integer k

$$P(X = k) = q^{k-1} p$$

where $q = 1 - p$.

Geometric random variables are used to model discrete "waiting time" phenomena. For example, consider an independent trials process with success probability on any given trial being p. If X is the number of the trial on which a success first occurs, then it is not hard to see that $P(X = k) = q^{k-1}p$. The reason for this is simply that in order for the first success to occur on the kth trial, it must be true that the first k–1 trials result in failure. Assuming that the process being observed is an independent trials process means that the probability of this happening is $q^{k-1}p$.

Example 4.1. Suppose that in a given floodplain, the probability of a flood in any given year is .2. Assuming that a structure has just been built in the floodplain, let's consider the random variable X that indicates the number of the year in which the structure is first subjected to a flood. Thus, saying $X = 1$ means that a flood occurs during the first year of existence of the structure, and so forth.

Assuming the occurrence of floods to form an independent trials process (some meteorologists might quibble with this assumption) makes X geometric with parameter $p = .2$. Thus the probability that the structure is first subjected to a flood in its fifth year is $P(X = 5) = .2 \times .8^4 = .08192$. It is implicit in the model here that a flood eventually *must* occur. This is seen by simply noting that the sum of the probabilities $P(X = 1) + P(X = 2) + \cdots = 1$. (See Problem 4.16.)

This is a discrete time model because time is being measured in terms of whole years rather than on a continuous scale. Clearly one could observe the same phenomena and choose to record time on a continuous rather than a discrete scale. To do so leads to consideration of a continuous rather than a discrete random

10, 3

variable. This alternate way of modeling the waiting time for a flood to occur will be considered in the next chapter.

One of the most interesting properties of geometric random variables is the "lack of memory" property stated in Proposition 4.2.

Proposition 4.2: If X is a geometric random variable, then

$$P(X = n + k \mid X > n) = P(X = k)$$

for all positive integers k and n.

Proof: Suppose X is geometric with parameter p. Then

$$P(X = k) = p \times q^{k-1}$$

where $q = 1 - p$. For a given positive integer n, it is easy to see that $P(X > n) = q^n$. The reason is that this probability is simply the sum of the infinite series

$$\begin{aligned} P(X = n+1) + P(X = n+2) + \cdots &= pq^n + pq^{n+1} + \cdots \\ &= pq^n(1 + q + q^2 + \cdots) \\ &= \frac{pq^n}{1-q} = \frac{pq^n}{p} = q^n \end{aligned}$$

An alternative and intuitively helpful way to observe that $P(X > n) = q^n$ is to think of X as modeling waiting time for success in an independent trials process. Then $P(X > n)$ is simply the probability that more than n trials are required, that is, that the first n trials result in failure. However, the probability that the first n trials result in failure is just q^n.

If $k > 0$ then the intersection of the events $(X > n)$ and $(X = n + k)$ is the event $(X = n + k)$, since $(X = n + k)$ is a subset of $(X > n)$. Thus

$$\begin{aligned} P(X = n + k \mid X > n) &= \frac{P(X = n + k)}{P(X > n)} = \frac{pq^{n+k-1}}{q^n} \\ &= pq^{k-1} = P(X = k) \end{aligned}$$

Let's interpret the meaning of this property in light of Example 4.1. Suppose in that example that a flood does not occur during the first five years. The conditional probability that the first flood occurs in year 12, based on this information, is then exactly the same as the initial (unconditional) probability that a flood would occur in year 7. In other words, there is no penalty for the 5 years of good luck. The conditional probabilities after the 5 years of good luck are the same as if we think of the process as starting all over again.

An example that is simpler still is the independent trials process consisting of repeated tosses of a coin. If we think of the random variable X that indicates the number of the trial on which the first head occurs, then X is geometric (with parameter $p = 1/2$ if the coin is unbiased). The lack of memory property says, for example, that if tails occur on the first two tosses, then the probability that the first head occurs on the fifth toss is the same as the *original* probability that the first head occurs on the third toss. In other words, you can forget about the first two tosses that have resulted in tails and think of the whole process as starting all over again. This mathematical model is consistent with most people's intuitive feeling that a coin doesn't have a memory, and it provides a little more evidence supporting the choice of an independent trials model as the "correct" one for repeated coin tossing.

Proposition 4.3: If X is a geometric random variable with parameter p, then

$$E(X) = \frac{1}{p}$$

Proof: By definition,

$$E(X) = \sum_{n=1}^{\infty} n\, P(X = n) = \sum_{n=1}^{\infty} n\, p\, q^{n-1} = p \sum_{n=1}^{\infty} n\, q^{n-1}$$

It is a pleasant fact from calculus that power series may be differentiated term by term. This means that if

$$f(x) = \sum_{n=0}^{\infty} a_n x^n$$

then

$$f'(x) = \sum_{n=1}^{\infty} n a_n x^{n-1}$$

In particular, if f is defined for $|x| < 1$ by the geometric series

$$f(x) = 1 + x + x^2 + \cdots = \frac{1}{1-x}$$

then

$$f'(x) = 1 + 2x + 3x^2 + \cdots = \frac{1}{(1-x)^2}$$

In particular, this implies that the infinite series that defines $E(X)$ in the first line of the proof does sum to $p \times (1 - q)^{-2} = p/p^2 = 1/p$.

Example 4.2. A quality-control engineer has possession of a box of devices. He is going to test the devices one at a time until he encounters one that does not meet design specifications. How long should this process take?

Solution: The easiest model here is an independent trials model. In such a model we would consider each device tested as an independent trial for which the results of the test are pass or fail (success or failure). It's not entirely clear that this model is perfect for this situation. For example, if the devices are aligned in the box in the order in which they come off the assembly line and if the assembly line goes through periods of malfunctioning, then it is possible that the "independence" required by an independent trials process might be lacking. For instance, if a given device in the box is bad, then it may be an indicator that there was something wrong on the assembly line as the device was going through, and in turn this might mean it is more likely that the others that came off the assembly line at about the same time are also bad.

On the other hand, it's also possible to visualize a situation where the defects are caused by some random phenomena, or where the devices have become all mixed up later, so that the testing of different devices does approximate independent trials. If an independent trials model is used, then the expected number of devices that would be tested in order to find a defective one is $E(X) = 1/p$, where p is the fraction of the devices that are defective. For example, if 1% are defective, the expected number that would have to be tested in order to find a defective one is then $E(X) = 100$.

Example 4.3. Proposition 1.1 in Chapter 1 is valid even when the sample space is partitioned into an infinite sequence of mutually exclusive events

$$A_1, A_2, A_3, \cdots$$

In that case we simply interpret the sums as infinite series. Let's look at a simple situation in which this idea is useful. Suppose a nickel and a dime are tossed simultaneously and that this experiment is repeated over and over. What is the probability that we will obtain a heads on the nickel *before* we obtain a heads on the dime, that is, that the first head on the nickel will appear on an earlier toss than the first on the dime?

Solution: Here we are actually interested in *two* quantities that correspond to geometric random variables: the number of the toss on which a head first appears on the nickel and the number of the toss on which a head first appears on the dime. Let's denote by X the number of the first toss when a head is obtained on the nickel, and by Y the number of the first toss in which a head is obtained on the dime. The problem then is to compute $P(X < Y)$. The events

$$(X = 1), (X = 2), \cdots$$

can play the roll of $A_1, A_2, \cdots$ in Proposition 1.1. Then from the multiplicative law of Proposition 1.1,

$$\begin{aligned} P(X < Y) &= P(X = 1)\, P(X < Y \mid X = 1) \;+\; P(X = 2)\, P(X < Y \mid X = 2) \;+\; \cdots \\ &= \left(\frac{1}{2}\right)\left(\frac{1}{2}\right) + \left(\frac{1}{4}\right)\left(\frac{1}{4}\right) + \left(\frac{1}{8}\right)\left(\frac{1}{8}\right) + \cdots \\ &= r + r^2 + r^3 + r^4 + \cdots \quad \text{where } r = \frac{1}{4} \\ &= r\,(1 + r + r^2 + \cdots) \\ &= \frac{1}{4} \times \frac{1}{1 - 1/4} = \frac{1}{4} \times \frac{1}{3/4} = \frac{1}{3} \end{aligned}$$

We have made use of the facts that

$$P(X < Y \mid X = 1) = \frac{1}{2}$$

$$P(X < Y \mid X = 2) = \frac{1}{4}$$

$$P(X < Y \mid X = 3) = \frac{1}{8}$$

and so on, in the derivation of the above series. Do you understand the basis for this? See Problem 4.17.

4.3 Discrete Uniform Random Variables

Suppose S is any set with n elements. A *uniform* probability measure on S is one which assigns the same probability to each element of S. So to make S into a uniform sample space all that is required is to assign probability $1/n$ to each element of S. This idea was heavily used in Chapter 1, and in Chapter 5 the close relation between this situation and the continuous uniform distribution (introduced in Chapter 3) will be examined.

Imagine the experiment of "choosing a random number from the interval [0, 1]." The continuous model for this would be to assume that the random variable X denoting the number chosen has as its density the uniform density on [0, 1].

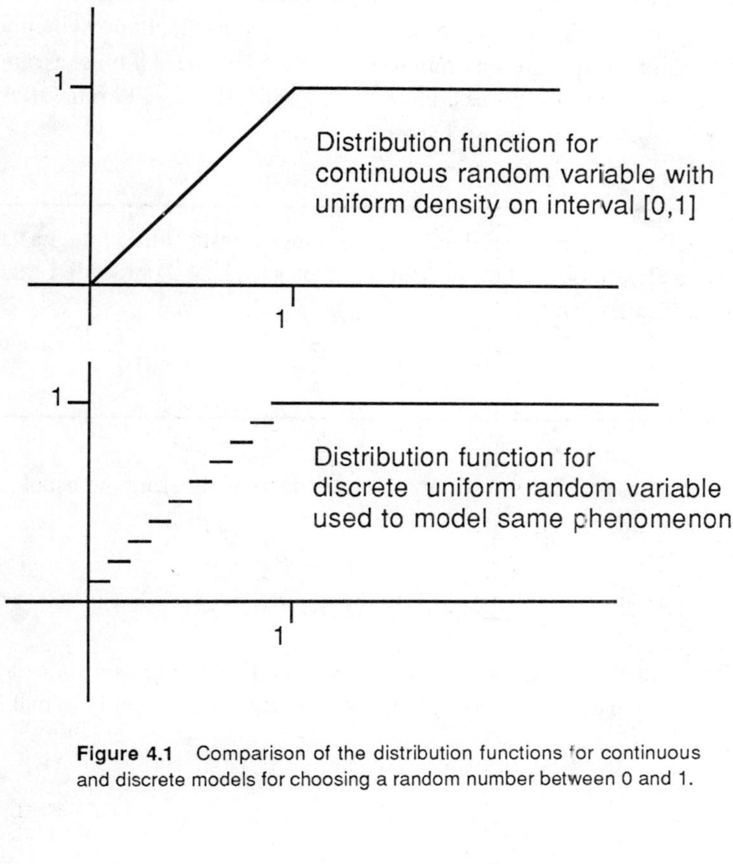

Figure 4.1 Comparison of the distribution functions for continuous and discrete models for choosing a random number between 0 and 1.

Suppose, however, that someone prefers to truncate the chosen number to a single decimal place. Let's denote by Y the truncated value. Thus, Y assumes values $0, .1, .2, \cdots, .9$ each with probability 1/10. It is highly instructive to compare the distribution function for Y, which is a discrete random variable, with that of the continuous random variable X. These are both shown in Figure 4.1.

4.4 Poisson Random Variables

The Poisson distribution is less easily motivated than those considered to this point. It is commonly used to model situations that involve some kind of random phenomena such as malfunctions of equipment, calls coming in to a switchboard, cars entering a parking lot, etc. In each of these situations, the Poisson distribution is used to model the number of times the phenomena occurs during a fixed time interval. For example, the number of calls received at a telephone switchboard during a 30-minute time interval might be assumed to be a Poisson random variable. A more thorough discussion of these applications will come after the discussion of Poisson processes in Chapter 7.

Definition 4.3: A random variable X that assumes non-negative integer values is said to have a *Poisson* distribution with parameter λ ($\lambda > 0$) provided that for each non-negative integer k

$$P(X = k) = e^{-\lambda} \frac{\lambda^k}{k!}$$

To note that these probabilities sum to 1 requires remembering the usual series expansion for the exponential function:

$$e^x = \sum_{n=0}^{\infty} \frac{x^n}{n!} \quad \text{for all real numbers } x \tag{4.1}$$

Replacing the x in this expansion by λ, we can see that the reason that the $e^{-\lambda}$ factor is present in the definition of the Poisson mass function is simply to make the probabilities sum to 1.

Example 4.4. Suppose the number of cars entering a certain parking lot during a 30-second time period is known to be a random variable having a Poisson mass function with parameter $\lambda = 5$. What is the probability that during a given 30 second period exactly 7 cars will enter the lot? What is the probability that more than 5 cars enter the lot during this time period?

Solution: $P(X = 7) = e^{-5} \times 5^7/7! = .104445$. Note that Problem 4.9 shows that in this example the average number of cars entering the parking lot during a 30-second time interval is also equal to 5; that is, the expected value is the same number as the parameter λ that appears in the mass function.

$P(X > 5) = 1 - P(X = 0) - P(X = 1) - \cdots - P(X = 5)$. Each of these terms on the right side of the equation is computed in the same way that $P(X = 7)$ was just computed.

A further consideration of Example 4.4 gives more insight. Suppose it is known that during a time period of 100 minutes, exactly 1000 cars entered the parking lot. (This is again an average of 5 cars every 30 seconds.) A particular subinterval of 30 seconds duration constitutes 1/200 of the total time. So we might think of the 1000 cars as 1000 independent trials, with each car having probability 1/200 of entering the lot during the given 30-second subinterval. From this point of view, the number of cars entering during the 30-second subinterval would be a binomial random variable with parameters $n = 1{,}000$ and $p = .005$. In that case, the probability $P(X = 7)$ would be

$$C(1000, 7)\,.005^7\,.995^{993} = .104602$$

Notice how close this answer is to the computation in Example 4.4 where an apparently quite different assumption is made about the probability distribution of the random variable in question. Numerically what is happening is summarized by saying that if n is "large" and p is "small," then the binomial distribution with parameters n and p is approximately equal to the Poisson distribution with parameter $\lambda = np$. A good understanding of precisely why this is so will have to wait until a discussion of Poisson processes in Chapter 7.

Proposition 4.4: If X is a Poisson random variable with parameter λ, then

1. $E(X) = \lambda$ and

2. $\sigma_X^2 = \lambda$

The proof of the first part of Proposition 4.4 depends only on the series representation for the exponential function, Equation 4.1. (See Problem 4.9.) The derivation of the variance of the Poisson distribution requires more manipulation with power series and is a more difficult computation. (See Problem 4.18.)

4.5 Hypergeometric random variables

Suppose 10 items of type A and 20 items of type B have become all mixed up, and 7 items from the total are to be chosen at random. What is the probability that 3 will be of type A and 4 of type B?

This question can be treated with the combinations formula (Proposition 1.2) from Chapter 1. The number of possibilities when 7 items are chosen from the total of 30 items is $C(30, 7)$. If 3 are of type A, this could occur in $C(10, 3)$ different ways. The 4 of type B can be any 4 of the 20 of type B, and there are $C(20, 4)$ different 4-element subsets of set B. Therefore, the number of ways of choosing 3 of type A and 4 of type B is the product $C(10, 3)\ C(20, 4)$. The probability of obtaining this distribution of items then, when 7 are randomly selected, is

$$\frac{C(10, 3)\ C(20, 4)}{C(30, 7)}$$

This is an example of the *hypergeometric distribution*. It arises when sampling is performed from a batch of objects that have been grouped into categories.

Definition 4.4: A random variable X has a *hypergeometric* distribution with parameters n, M, and N (positive integers with $n \le N$ and $M \le N$) if

$$P(X = k) = \frac{C(M, k)\ C(N - M, n - k)}{C(N, n)}$$

Familiarity with the hypergeometric distribution is useful in understanding the two basic kinds of sampling. These two sampling methods can be described as *sampling with replacement* and *sampling without replacement*. Specifically, the hypergeometric distribution arises in sampling without replacement. In the language of sampling theory, the parameters in Definition 4.4 are as follows: n is the sample size, N is the population size, and M is the number of items in the total population of a particular type, say type A. The random variable X then corresponds to the number of type A items that are chosen when n items are chosen from a total population of N objects. The rationale is basically the same as that used in the example that precedes the definition. The denominator is the number of possibilities for the sample drawn, n items being drawn from the population of N objects. $C(M, k)$ is the number of ways that k items could be chosen from the M items of type A. And $C(N - M, n - k)$ is the number of ways that the other $n - k$ items in the sample could come from the $N - M$ items that are *not* of type A.

Example 4.5. Of the items produced on an assembly line, 5% are defective. From a group of 17 items taken from the line, what is the probability that exactly 2 of the 17 will be defective?

Solution: We will assume an independent trials model. (Any other approach would require more information than we are given.) Assuming the 17 items to be independently produced means that the probability of 2 defects will be viewed as the probability of 2 successes in 17 trials where $p = .05$. This would give

$$P(X = 2) = C(17, 2)\ .05^2\ .95^{15} = .1575$$

where X is a binomial random variable with parameters $n = 17$ and $p = .05$.

Example 4.6. Assume now that 17 items are to be tested from a batch of 100 items of which 5% are defective. What is the probability that exactly 2 in the sample of 17 will be defective?

Solution: The question that has to be decided first is this: What are we going to do with an item after we test it? Are we going to lay it aside, or are we going to put it back in with the others so that the same item might get tested again? (In the latter case we would not be testing 17 *different* items necessarily.)

First let's suppose that we lay items aside once they are tested, so that there is no chance of testing the same item twice. In this case 17 items are being chosen from the batch of 100, of which 95 are good and 5 defective. If X is the number of defectives in the sample of 17, then X is a random variable having a hypergeometric distribution with $N = 100$, $n = 17$, and $M = 5$. And

$$P(X = 2) = \frac{C(5, 2)\ C(95, 15)}{C(100, 17)} = .1660$$

This constitutes sampling *without* replacement.

Suppose that we *do* replace each item after testing. In that case each item tested can be any of the 100 items, of which 5 are defective. The results of any given test are independent of previous tests, since tested items have been replaced in the population. This puts us back in an independent trials situation, and so the probability of 2 defects among the 17 tests would again be

$$C(17, 2)\, .05^2\, .95^{15} = .1575$$

as in Example 4.5. This is an example of *sampling with replacement. Sampling with replacement leads to the binomial distribution, whereas sampling without replacement leads to the hypergeometric distribution.*

4.6 Probabilities Based on Observed Data

Often it is useful to study the distribution of values in some fixed data set. The following example illustrates the relation between this idea and the concept of a probability distribution. One can, of course, divorce the data set analysis from probability altogether, but the idea of a *histogram*, or *frequency distribution*, is in fact quite similar to the ideas of the probability mass function and the density function.

Example 4.7. Failure data is being maintained for a particular type of equipment. For 250 items that have been in service for a year or more, the manufacturer has compiled the data which appears in the table below.

If we wish to assign probabilities for number of failures, a direct method of doing so is obvious. Since 93 of the 250 pieces of equipment experienced no failure during the year, we would say that the probability of no failure *based on this data* is 93/250 = .372. This leads to a sample space consisting of 7 elements with nonzero probability, specifically $S = \{0, 1, 2, 3, 4, 6, 8\}$. The probability of 8 failures in a year is only 1/250 = .004.

No. of failures	Observed frequency
0	93
1	98
2	44
3	11
4	2
5	0

6	1
7	0
8	1

There are two important observations to make about this simple situation. One is that the decision to include or exclude the outcomes 5 and 7 shown in the table is arbitrary. If they are included, they will be assigned probability 0. However, having them included with probability 0 serves no useful purpose. Rather than to include them in the sample space with assigned probability 0, it is simpler to leave them out. In reality, there is no more reason to include the outcomes 5 and 7 in the sample space with assigned probability zero than there is to include the values 9 and 10 or any other numbers. The only thing special about 5 and 7 is that they happen to be included in the table of data above.

The second important point in this example is that a probability model based on this data does not necessarily have predictive value for other data. For example, if this data were accumulated for items made at one plant, it does not mean necessarily that similar equipment made at another plant would have the same failure rates. Nor does it mean that equipment from the *same* plant made during a different time period would have similar failure rates. The fact that 5 or 7 failures were never observed during the period that the table covers does not mean that the next piece of equipment made at the plant will not fail 5 or 7 times during the first year.

Different data sets may or may not reflect a similar distribution. An analysis of the "correlation" between data sets is a question for statistics. The important probabilistic concept is that the probabilities introduced in this example to describe this data set do only precisely that: *The probabilities describe the probability distribution of the given data set.*

The following figure is a *histogram*. Histograms are frequently used to display frequency distributions. In the figure the values 5 and 7 are shown only to maintain the linearity of the horizontal scale. Notice that the idea of a histogram is basically the same as the idea of displaying the probabilities of the various outcomes. If the values on the vertical scale were divided by 250, the heights of the bars would be the probabilities of the outcomes. This would then be a visual representation of the probability mass function of the random variable that gives the number of failures during the first year of operation for the equipment described by the data set. (This would not technically be a "graph" of the probability mass function because the mass function satisfies $p(x) = 0$ except for a few integer values of x. However, it could be construed as the graph of the density function of a closely related continuous random variable. See Problem 4.15.)

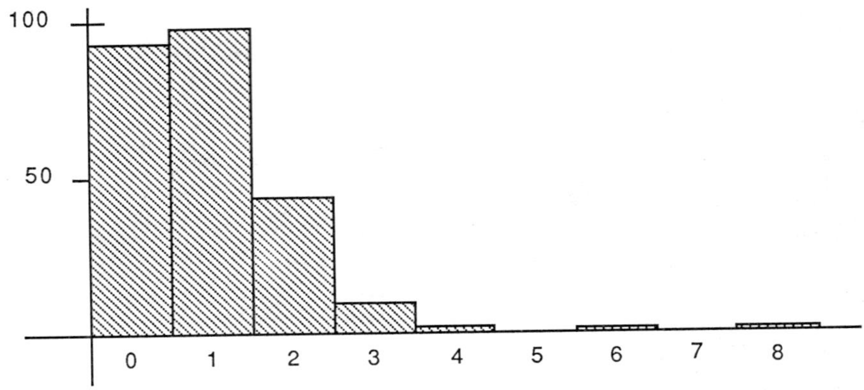

Problems

4.1 Three cards are drawn without replacement from the 13 spades in a deck of cards. Let X denote the number of face cards drawn. (The ace, king, queen, and jack are the face cards.) Determine the probability mass function for X.

4.2 If a coin is tossed repeatedly, what is the probability

(a) that the first head occurs after the 5th toss?

(b) that exactly 3 heads occur in the first 10 tosses?

4.3 A random variable X assumes values 1, 2, 3, $\cdots$, 100, and for each integer k where $1 \leq k \leq 100$, $P(X = k) = .01$. Using Σ notation, give a sum that would evaluate each of the following:

(a) $E(X)$ (b) var (X) (c) $E(Y)$ where $Y = 2^X$

4.4 A coin is tossed 3 times. Let X denote the number of heads obtained. Let Y be the absolute value of the number of heads minus the number of tails. Compute $E(Y)$ in each of the following ways:

(a) Write Y as a function of X; that is, determine g such that $Y = g(X)$ and then use Proposition 3.6.

(b) Determine the probability mass function for Y and compute $E(Y)$ directly from Definition 3.6.

4.5 Suppose $S = \{w_1, \cdots, w_n\}$ is a uniform sample space, and suppose X is a random variable defined on S. For $1 \le k \le n$, let $x_n = X(w_n)$. Show that $E(X)$ is simply the arithmetic average of the numbers $x_1, \cdots, x_n$, that is, show that

$$E(X) = \frac{1}{n} \times (x_1 + \cdots + x_n)$$

4.6 Suppose X is a Poisson random variable with parameter λ. What is the "most probable" value for X to assume? In other words, what value of n has probability such that $P(X = n) > P(X = k)$ for all other positive integers $n \neq k$? Show that if λ is itself an integer, then there is a tie for the "most probable value" of X.

4.7 Suppose that it is known that the number of components to fail in a complex electrical device during a one-day time period is a random variable X having a Poisson distribution with parameter $\lambda = 5$. What is the probability that there will be exactly 3 failures next Tuesday?

4.8 Suppose it is known that a particular electrical device had 50 components fail during the past 10 days. What is the probability that exactly 3 should have failed yesterday? (The model you should use is that of an independent trials process. Each of the 50 breakdowns could have occurred yesterday or on 1 of the other 9 days. Considering the days equally likely means that a particular breakdown had a 1/10 chance of occurring yesterday and a 9/10 chance of occurring one of the other days.)

Notice that the answers to this problem and to Problem 4.7 are almost identical. In each case we are looking at a random variable for which $E(X) = 5$. (For the Poisson case, this is a consequence of Problem 4.9.) In fact there is a very close relationship between the Poisson and the binomial distributions which these problems illustrate. This relationship will be explored in Chapter 7.

4.9 Show that if X is a Poisson random variable with parameter λ, then $E(X) = \lambda$. [Hint: This is easy. Just write out the series that defines $E(X)$ and make use of Equation 4.1.]

4.10 Determine the variance of the discrete uniform random variable that assumes values 1, 2, 3, 4, and 5, each with probability .2.

4.11 Generalize Problem 4.10 as follows: Suppose X is a discrete uniform random variable assuming values $x_1, \cdots, x_n$, each with probability $1/n$. Use the expression for var(X) given in Proposition 3.5 to derive an expression for the variance of X. (Use Σ notation.)

4.12 A random variable X gives the voltage output from an acoustical transducer. The voltage varies from –3.5 to +3.5 volts, but X measures the voltage as rounded off to the nearest integer. Assuming that the probability distribution for X is the discrete uniform distribution and that X assumes the values $\{-3, -2, -1, 0, 1, 2, 3\}$, find the cumulative distribution function for X and sketch its graph.

4.13 A "black box" transmits binary data in the form of 0's and 1's which go into the box and are transmitted out according to the following probabilities. Let's denote by X and Y the random variables that give the input digit and the output digit, respectively. The relation between X and Y is as follows: $P(Y = 0 \mid X = 0) = p_0$ and $P(Y = 1 \mid X = 0) = 1 - p_0$. On the other hand, if a 1 is input then the probability is p_1 that a 1 is output and $1 - p_1$ that a 0 is output.

(a) If the probability is 1/3 that a given input digit is 0 and 2/3 that it is 1, find the probability that the output digit is a 0 and the probability that it is a 1. These will be expressed in terms of p_0 and p_1.

(b) If an output digit 1 is observed, what is the probability that the input digit was a 0? (This is a conditional probability question. Continue to assume that the unconditional probability of 0 as the input digit is 1/3.)

4.14 The number of letters mailed out from a business office in a day was surveyed over a period of time. Results of the survey were as follows:

No. of letters	No. of days observed
0	3
1	4
2	9
3	14

4	12
5	9
6	5
7	5
8	4
9	1
10 or more	4

Construct a probability distribution based on this data with the outcome "10 or more" considered as one of the elements of the sample space. Construct a histogram for this data set, and sketch a graph of the distribution function of the random variable that corresponds to the number of letters leaving the office in a day.

4.15 A discrete random variable X assumes values 1, 2, and 3 with probabilities 1/2, 1/3, and 1/6, respectively. Do you see a relation between X and the continuous random variable Y having density function equal to the step function that takes on values 1/2, 1/3, and 1/6 and whose graph is shown below?

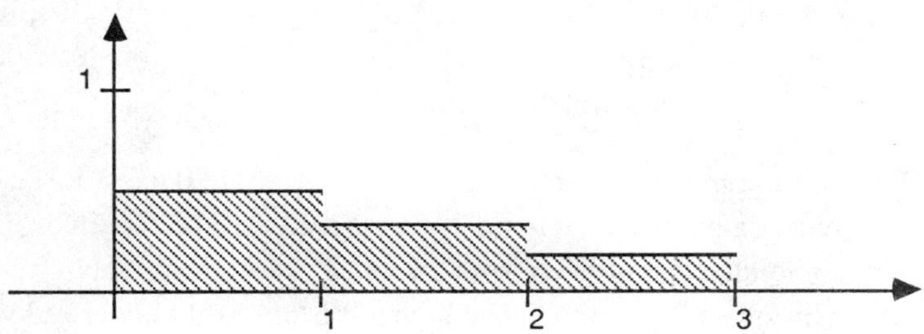

Specifically, show that $F_Y(t) = F_X(t)$ if t is an integer. Now compare the graphs of F_Y and F_X.

4.16 Use the idea of a geometric series to show that for any geometric random variable X, $P(X = 1) + P(X = 2) + P(X = 3) + \cdots = 1$.

4.17 In Example 4.3, give an explanation for why $P(X < Y \mid X = 1) = 1/2$, $P(X < Y \mid X = 2) = 1/4$, and $P(X < Y \mid X = 3) = 1/8$, and so on.

4.18 Show that if X is a Poisson random variable with parameter λ, then

$$\sigma_X^2 = \lambda$$

[Note: This problem should be attempted only if you are comfortable with

$E(X^2)$. One way to do this is to write $E(X^2)$ in the form $\lambda e^{-\lambda}g(\lambda)$, where $g(\lambda)$ is the sum of a power series. While it is not possible to evaluate the series defining $g(\lambda)$ directly, the sum of the series can be found by looking instead at the antiderivative of $g(\lambda)$, by applying Equation 4.1 to this series, and then differentiating to obtain $g(\lambda)$.]

4.19 A motel owner has bought 7 television sets from UltraView TV Corporation. If 26% of UltraView's televisions have to be returned for repair during the first year of operation, what is the probability that the motel owner will have to return more than two of his sets? (Make sure that you know how to frame your solution in terms of the binomial distribution.)

4.20 The motel owner of Problem 4.19 put 4 of his 7 televisions in private rooms and 3 in the lobby. If 3 of them have to be returned for repairs during the first year, what is the probability that 2 of the 3 to be repaired are ones that were in the lobby? [Hint: Assume that the seven televisions are all equally likely to fail. Frame your model here in terms of the hypergeometric distribution. What are n, M, and N? Can you summarize explicitly what assumption you are making about this question when you decide to model it using the hypergeometric distribution?]

4.21 Suppose the motel owner of Problems 4.19 and 4.20 has 3 instances during the year of a guest damaging a TV so that it must be repaired. What is the probability that two of the repairs are to televisions in the lobby?

This question is ambiguous in that it is not explicitly stated whether the same television may be damaged more than once. Common sense would suggest that it might be; so let's assume that each case of vandalism is equally likely to happen to any of the 7 televisions. Which of the probability distributions discussed in this chapter is your model then based on? What is the relation between this problem and Problem 4.20?

4.22 Suppose you were going to sample the opinion of 200 registered voters in the United States as a gauge of public opinion on some issue. Would it make a significant difference whether you chose sampling with replacement or sampling without replacement? Why or why not?

Chapter 5: Continuous Models

Continuous random variables arise when quantities are being observed that may assume values throughout some interval of numbers. The necessary information to do computations involving such random variables is carried by the density function. In Chapter 3 we saw a few elementary examples of density functions, and in this chapter we will see how some of the most common are used to model real-world phenomena. Just as was the case with discrete models, the skill that the modeler must bring to a problem involving continuous random variables is the experience and intuitive understanding necessary to judge what kind of distribution "fits" the situation.

5.1 Continuous Uniform Random Variables

We introduced the uniform density on [0, 1] in Chapter 3. This concept is just as natural for an arbitrary interval. Keep in mind that we will have to adjust the value of the function in order to keep the area under the graph equal to 1.

Definition 5.1: The *uniform* density on an interval $[a,b]$ is the function defined by

$$f(t) = \begin{cases} \dfrac{1}{b-a} & \text{if } a \leq t \leq b \\ 0 & \text{otherwise} \end{cases}$$

A random variable that has this density function is said to be *uniform* (or *uniformly distributed*) on the interval $[a, b]$.

The reason $f(t)$ must assume the value $1/(b-a)$ on the interval $[a, b]$, of course, is that the area under the graph is required to be 1, in accordance with Definition 3.3. It should be apparent from the symmetry of this density function that the mean of a uniform random variable is simply the midpoint of the interval and that the variance is greater the longer the interval. (See Problem 5.1.)

Figure 5.1 shows two uniform density functions, the uniform density on [0, 1/2] and the uniform density on [3, 5]. The area under the graph (which is required to equal 1) is shaded in each instance.

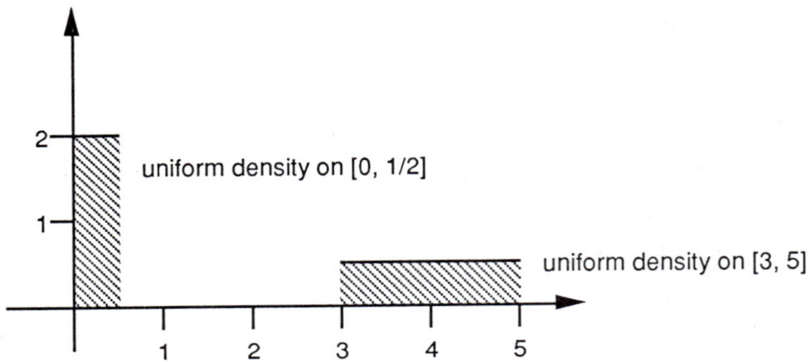

Figure 5.1 Two uniform density functions.

5.2 Exponential Random Variables

Definition 5.2: A random variable X is said to be *exponential* with parameter λ provided that X has density function

$$f(t) = \begin{cases} \lambda e^{-\lambda t} & \text{for } t \geq 0 \\ 0 & \text{for } t < 0 \end{cases}$$

Here λ is required to be a positive constant in order to make $f(t)$ integrable.

Exponential random variables were encountered in Chapter 3, for example in Problems 3.13, 3.14, and 3.15. The effect that the specific value of λ has on the shape of the density function is illustrated by Figure 5.2. When λ is large, the associated random variable will have a high probability of assuming a value near 0. When λ is small, the values assumed by the random variable are much more spread out. Figure 5.2 shows two scenarios corresponding to $\lambda = 2$ and $\lambda = 1/2$.

If X is an exponential random variable with parameter λ, then $E(X) = 1/\lambda$ (see Problem 3.15). This characteristic, that the expected value is the reciprocal of the parameter in the distribution, is shared with the discrete geometric distribution; if a discrete random variable X is geometric with parameter p, then $E(X) = 1/p$ (Proposition 4.3). In fact, the similarity between geometric and exponential random variables is substantial. The exponential distribution shares the same lack of memory property as does the geometric, and the exponential distribution is used to model continuous "waiting time" situations in the same way that the geometric is used to measure discrete "waiting time" for success in an independent trials process. This application of the exponential distribution will be treated in depth in the investigation of Poisson processes in Chapter 7.

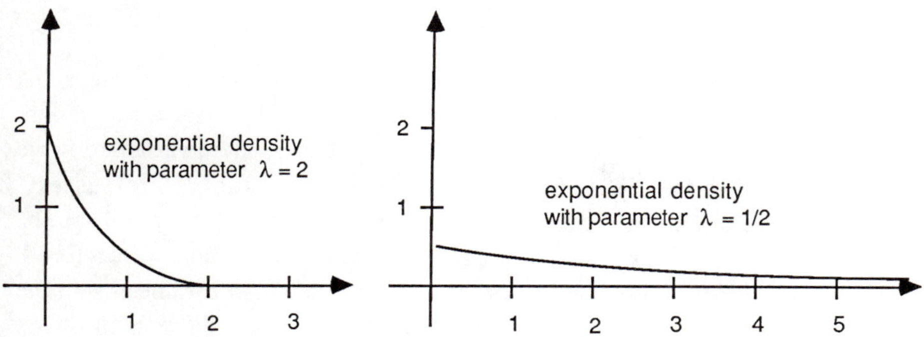

Figure 5.2 Two exponential density functions.

Proposition 5.1: If X is an exponential random variable, then for any positive numbers s and t,

$$P(X > s + t \mid X > t) = P(X > s)$$

lack of memory

Proof: First observe that for any positive number w,

$$P(X > w) = \lambda \int_w^\infty e^{-\lambda t}\, dt = e^{-\lambda w}$$

(This computation has appeared previously as Problem 3.14.) From this, it follows that $P(X > s + t) = e^{-\lambda(s+t)}$, and so

$$P(X > s + t \mid X > t) = \frac{e^{-\lambda(s+t)}}{e^{-\lambda t}} = e^{-\lambda s} = P(X > s)$$

Example 5.1. Suppose that time to first failure of an electrical component is an exponential random variable with expected value 25 months. If X denotes the random variable that measures time to failure and if X is assumed to have an exponential density function, then the knowledge that $E(X) = 25$ months requires that the parameter λ in the distribution be $\lambda = .04$ (so that $1/\lambda$ will be 25).

In the context of these assumptions, what is the probability that the device lasts longer than 30 months?

Solution: From Problem 3.14, if $s > 0$ then $P(X > s) = e^{-\lambda s}$. Thus

$$P(X > 30) = e^{-30 \times .04} = e^{-1.2} = .30119$$

Now by way of comparison, consider a discrete model for the same experiment. Suppose someone wants to measure the month in which first failure occurs as an integer value. Let Y be this integer-valued random variable. If the sequence of months is considered to be independent trials in which "success" means that the device fails and "failure" means it doesn't, then Y will be the number of the month in which the first failure of the device occurs. Therefore Y will have a geometric distribution, and if $E(Y) = 25$ then the parameter p in the geometric distribution must be $p = .04$. So $P(Y > 30) = (1 - p)^{30}$ (as observed in the proof of Proposition 4.2) $= .96^{30} = .29385$. (One could, of course, identify the failure of the device with the outcome called "failure" in the independent trials process. While the wording seems more natural this way, we would then be considering "waiting time for failure" rather than "waiting time for first success," which was the terminology we used in discussing geometric random variables. It should be clear that the differences are purely linguistic and have no effect on the underlying mathematics.)

In the continuous model for this "waiting time" experiment, $P(X > 30)$ and $P(X \geq 30)$ would of course be the same since $P(X = 30) = 0$. In the discrete model, however, $P(Y \geq 30) = P(Y > 29) = .96^{29} = .30610$. Notice that when the discrete and continuous models are compared, the discrete model yields a slightly higher or slightly lower probability than does the continuous model, depending on whether we interpret the question as being "what is the probability that the first failure occurs after month 30" or as being "what is the probability that the first failure occurs during or after month number 30?"

You might be interested to know that Proposition 5.1 characterizes exponential random variables. In other words, a continuous random variable that has the lack of memory property exhibited in Proposition 5.1 necessarily has an exponential density function.

5.3 Normal Random Variables

The normal distribution is widely used in statistics and plays an important theoretical role in probability. This distribution is commonly used to model phenomena in which a continuous random variable assumes values symmetrically weighted around some central value which is the mean of the random variable. Whereas the exponential density functions have only one parameter (referred to as λ in Definition 5.2), there are two parameters in normal density functions. The shape of normal density functions is that of "bell-shaped" curves. The location of the curve along the horizontal axis is determined by the mean, and the "steepness" of the bell is determined by the variance.

Before considering the most general types of normal density functions, we must first understand the *standard* normal density function.

Definition 5.3: The *standard normal* density function is the function

$$f(t) = \frac{1}{\sqrt{2\pi}} e^{-t^2/2}$$

A random variable X that has this density function is called a *standard normal random variable.*

It is not elementary to show that the standard normal density is a density function at all. To show that

$$\int_{-\infty}^{\infty} e^{-t^2/2}\, dt = \sqrt{2\pi}$$

requires a special trick from multivariable calculus. Finding the mean and the variance, however, are straightforward. Proof of the Proposition 5.2, which follows, is left as Problem 5.19.

Proposition 5.2: If X is standard normal, then $E(X) = 0$ and $\text{var}(X) = 1$.

More general normal random variables have density functions of the following type.

Definition 5.4: The *normal density function with parameters* μ *and* σ (where $\sigma > 0$) is the function

$$f(t) = \frac{1}{\sigma\sqrt{2\pi}} e^{-(t-\mu)^2/2\sigma^2}$$

If a random variable has this density function, it is said to be a *normal random variable with parameters* μ *and* σ.

Unfortunately, the antiderivatives of the normal density functions cannot be expressed in terms of known elementary functions. This means that one cannot do computations involving normal density functions and evaluate the integrals using the familiar fundamental theorem of calculus, that is, by finding an antiderivative and evaluating at the endpoints of integration. Various numerical approximation methods can be used, and this is commonly done to determine values of the distribution function for a standard normal random variable. Further discussion of ways to do this (including the use of tables of values) appears later in this chapter. Fortunately, it is sufficient to have a method for evaluating the standard normal distribution function. Once the standard normal distribution function is known, Proposition 5.3 can be used to compute probabilities involving any normal random variable. The key idea behind the proposition is that every normal random variable is a linear function of a standard normal one. The proof of this proposition is based on the change of variables $s = (t - \mu)/\sigma$, which converts the normal density with parameters μ and σ to the standard normal density. (See Problem 5.20.)

For a normal density function, the parameter μ turns out to be the mean of the corresponding random variable and σ is the standard deviation. (See Proposition

5.3.) Thus the value of μ determines where the "bell" is located on the t axis. This is shown in Figure 5.3. It is important to realize that the general shape of the bell is determined by the parameter σ. A small value of σ produces a tall skinny bell, whereas a large value of σ causes the values to be much more spread out with a correspondingly lower peak.

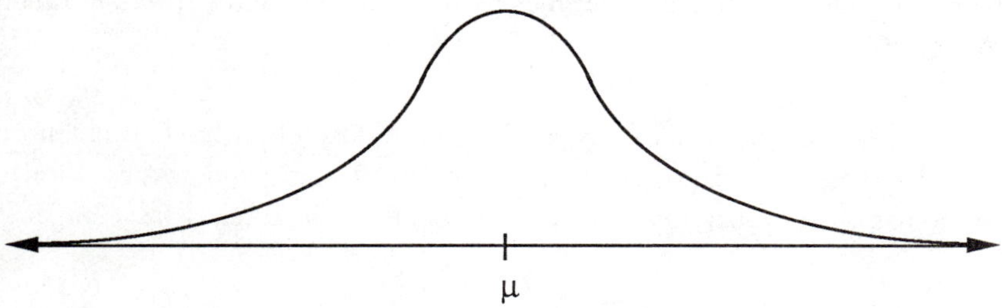

Figure 5.3 The graph of a typical normal density function.

The tremendous help that Proposition 5.3 provides is that it enables us to convert problems about arbitrary normal random variables into problems about standard normal random variables. This means that we can frame all computations in terms of the standard normal distribution function.

Proposition 5.3:

1. If X is normal with parameters μ and σ, then X has mean μ and variance σ^2.
2. If X is normal with parameters μ and σ, then the random variable $(X-\mu)/\sigma$ is standard normal.
3. If X is standard normal and μ and σ are numbers with $\sigma > 0$, then the random variable $Y = \sigma X + \mu$ is normal with parameters μ and σ.

Comment: Remember that for any random variable X,

$$P(a < X \leq b) = F_X(b) - F_X(a)$$

This says that all probabilities of interest can be computed once the distribution function F_X is known. Since all normal random variables can be expressed as a function of a standard normal one (in light of part 2 of Proposition 5.3), this means that values of the standard normal distribution function are all that is needed. Because of the importance of the standard normal distribution function, it is common to devote a special symbol to serve as a name for it. Often the symbol Φ is used to denote the standard normal distribution function, and we will follow that convention. (Other symbols sometimes used for the normal distribution function are N_Z and Q.)

Example 5.2. A particular type of lightbulb has a lifetime that is a normal random variable with expected value 1,000 hours and standard deviation 200 hours. What is the probability that a lightbulb of this type actually lasts longer than 900 hours?

Solution: Let X denote the lifetime of the bulb. Then X is normal with parameters $\mu = 1{,}000$ and $\sigma = 200$. Furthermore, if we let Y denote $(X - \mu)/\sigma$, then Y is standard normal.

We wish to know $P(X > 900)$. However,

$$\begin{aligned} P(X > 900) &= P(\,(X - \mu)/\sigma > (900 - \mu)/\sigma\,) \\ &= P(Y > -.5) = 1 - P(Y \leq -.5) \\ &= 1 - F_Y(-.5) = 1 - .3085 = .6915 \end{aligned}$$

Thus the probability that the bulb lasts more than 900 hours is approximately .6915. (See Problem 5.14.)

Why would we choose a normal distribution to model the lifetime of a lightbulb? There is at least one logical inconsistency in doing so. It is that all normal density functions are positive for all real values of t, meaning that it is theoretically possible for any normal random variable to assume a value in any interval whatsoever. To assume a normal distribution with mean $\mu = 1{,}000$ and standard deviation $\sigma = 200$ means we are necessarily going to come up with a positive probability that the lightbulb has a *negative* lifetime! While this may seem troubling at first, it is of little practical significance. Remember that most mathematical models are only approximate. If you actually compute $P(X < 0)$, where X is a normal random variable with mean $\mu = 1{,}000$ and standard deviation $\sigma = 200$, you will find that the probability is negligibly small. In fact, a normal distribution may be quite accurate for such a device if indeed the manufacturing process is consistent enough that observed lifetimes do tend to cluster around some mean value.

Both the normal and the exponential distributions are used to model "time to failure" for various devices. The normal distribution fits situations such as the lightbulb example above where some kind of actual "wearing out" or "burning out" is taking place. Clearly, a bulb that has burned for 1,000 hours is not physically the same as a new bulb. There is no "lack of memory" property in this situation. On the other hand, something like a piece of electrical cable may not be undergoing any physical deterioration during use and may be no more likely to fail after 10 years of use than the day it was installed. Instead, it might fail only because of some random occurrence having nothing to do with any aging process, and thus its lifetime might well have the "lack of memory" property exhibited by the exponential distribution.

Example 5.3. Even such a clearly discrete phenomenon as coin tossing is related to the normal distribution. In fact, if X is any binomial random variable for which the parameter n is "large," then $(X - \mu_X)/\sigma_X$ is "almost" standard normal if μ_X and σ_X denote the expected value and the standard deviation, respectively, for X. This fact, Proposition 5.4, shows a very important way the normal distribution can be used in approximations.

Let's compute, for example, the approximate probability that in 1,600 tosses of a fair coin, less than 775 heads occur. Let X denote the number of heads in 1,600 tosses. Then X is binomial with $n = 1{,}600$ and $p = .5$. From Proposition 4.1, we know that the values of μ_X and σ_X are $\mu_X = 800$ and $\sigma_X = \sqrt{npq} = 20$. Therefore,

$$Y = \frac{X - \mu_X}{\sigma_X} = \frac{X - 800}{20}$$

is "approximately" standard normal. Thus,

$$\begin{aligned} P(X < 775) &= P(X - 800 < -25) \\ &= P[(X - 800)/20 < -1.25] \\ &= P(Y < -1.25) \\ &= F_Y(-1.25) \approx \Phi(-1.25) \approx .1056 \end{aligned}$$

Some ways of obtaining this value of $\Phi(-1.25)$ are discussed in the next section. All require use of some kind of numerical method.

What we have done in this example is to use the normal distribution as an approximation to the binomial. If great accuracy is desired in such a calculation, there is a trick called the *continuity correction* that can be brought into play. Recall that our original task was to determine $P(X < 775)$. Since the number of heads in reality can assume only integer values, this is the same as $P(X \leq 774)$. However, once we move to the normal distribution, it makes no difference whether we use $<$ or $\leq$ in the inequality. (Remember that for continuous random variables, endpoints may be disregarded.) So, should we use 775 or 774? The result of using one is that our approximation is a little too high, and with the other it is a little too low. To compromise and use 774.5 would give a more accurate result, though the result already obtained, .1029, is quite good. Problem 5.29 asks you to redo the calculation using 774.5 rather than 775 and to observe how much this changes the answer. Problem 5.16 shows that the continuity correction can provide a substantial improvement when the parameter n appearing in the binomial distribution is relatively small.

The method used in Example 5.3 is based on Proposition 5.4. This is a special case of an important theoretical result called the central limit theorem. A more general form of the theorem is stated in Chapter 6.

Proposition 5.4: If X is binomial with parameters n and p and if n is "large," then the random variable

$$Y = \frac{X - np}{\sqrt{npq}}$$

is approximately standard normal. (In other words, the distribution function of Y is approximately the same as the standard normal distribution function if n is large.)

Bear in mind that the larger n is, the better the approximation. This is fortunate, because with large values of n, the binomial distribution is cumbersome to use. So the approximation provided by the normal distribution is most accurate just where it is most needed, for large values of n.

The relation between the binomial and the normal distribution is of more than just theoretical interest. It provides a useful tool for estimating the reliability of "sampling" strategies. The following example illustrates this.

Example 5.4. A pharmaceutical firm wishes to test the effectiveness of a new drug. In each patient, the drug either will or will not achieve a desired result. Different patients being administered the drug may be viewed as independent trials of an experiment with success probability p, where p is the probability that the drug will have the desired effect.

The problem, of course, is that the value of p is not known. The company wishes to conduct a test to estimate p. The idea of the test is simple. Since probabilities should reflect a theoretical "relative frequency" of occurrence, a large number of patients can be tested with the drug and the fraction on whom the drug works should then approximate p. A strategy, though, is needed in order to decide how many patients the drug should be tested on.

What the company would like in the end is a conclusion similar to this: "Our drug has the desired effect on 84% of all patients, plus or minus 3%. Furthermore, the probability that this claim is correct is at least 95%."

What kind of mathematical basis could there be for such a claim?

Solution: Suppose the drug is tested on n people, assumed to be independent trials, and that the probability that the drug works on any given person is p (unknown). The number of people on whom the drug has the desired effect then is a binomial random variable with parameters n and p. We will denote this random variable by X.

The fraction of people on whom the drug works is then X/n, and this is what will be used to estimate p. Specifically, let's concentrate on the question as to how likely it is that this approximation is not within .03 accuracy. In other words, let's look at

$$P\left(\left| \frac{X}{n} - p \right| > .03 \right)$$

Now

$$\begin{aligned} P\left(\left| \frac{X}{n} - p \right| > .03 \right) &= P\left(\frac{|X - np|}{n} > .03 \right) \\ &= P\left(\frac{|X - np|}{\sqrt{npq}} > \frac{.03\sqrt{n}}{\sqrt{pq}} \right) \\ &\leq P\left(\frac{|X - np|}{\sqrt{npq}} > .06\sqrt{n} \right) \qquad (5.1) \end{aligned}$$

The reason this latter inequality is true depends on a little elementary calculus. Check that the function $f(p) = p(1 - p)$ takes on its maximum value at $p = 1/2$, where $f(p) = 1/4$. This means that pq, where $q = 1 - p$, is always less than or equal to 1/4, and therefore $\sqrt{pq} \le 1/2$ for all possible values of p and $q = 1 - p$. Therefore,

$$\frac{.03\sqrt{n}}{\sqrt{pq}} \ge .06\sqrt{n}$$

Because of this inequality, the event whose probability forms the left side of Inequality 5.1 above is a subset of the event whose probability is on the right. So the inequality is a special case of the fact that $P(A) \le P(B)$ if $A \subset B$.

Here's where the normal distribution comes in. Recall that the random variable

$$\frac{X - np}{\sqrt{npq}}$$

is approximately standard normal. Problem 5.30 asks you to show that if Z is a standard normal random variable, then $P(|Z| > d) = 2 - 2\Phi(d)$ for any positive number d. Therefore, the rightmost term in Inequality 5.1 is approximately equal to $2 - 2\Phi(.06\sqrt{n})$.

The company would like to be 95% confident of their claim, that is, the goal is that this term be no greater than .05. So let's just set $2 - 2\Phi(.06\sqrt{n})$ equal to .05 and solve for n. Then, since $2 - 2\Phi(.06\sqrt{n}) = .05$, we know $\Phi(.06\sqrt{n}) = .975$. In the next section we will discuss ways of evaluating the function Φ and its inverse. If $\Phi(.06\sqrt{n}) = .975$, then $.06\sqrt{n} \approx 1.96$ (use the table in Appendix B if you like), and solving for n gives $n \approx 1067$.

Conclusion: If the firm wishes to claim with 95% confidence that they have estimated the probability p to within .03 by testing n cases and then estimating p by the fraction of the cases on which the treatment works, it will be necessary for them to test at least 1067 patients with the drug.

Clearly there are numerous situations to which the techniques of the above example would be applicable. The most familiar is that of the public opinion research firms that make statements such as, "If the election were held today, candidate A would get 42% of the vote, plus or minus 3%." The 95% confidence level is so often used that it has become something of a standard and is often left unsaid in statements such as this.

There is another useful tool for estimating the probability that a random variable differs from its mean by more than a given amount. It is called

Chebyshev's inequality and appears as Proposition 5.5. The estimate that Chebyshev's inequality provides is necessarily going to be rather crude since the hypotheses are so weak. Nothing is assumed about the nature of the distribution of X other than that X has finite mean and variance. Nevertheless it is a powerful theoretical tool that is often useful for the qualitative rather than the quantitative information that it provides.

Proposition 5.5 (Chebyshev's Inequality): Suppose that X is a random variable having finite mean μ and variance σ^2. Then for any positive number ε

$$P(|X - \mu| \geq \varepsilon) \leq \frac{\sigma^2}{\varepsilon^2}$$

When expressed in terms of the complementary event this says

$$P(|X - \mu| < \varepsilon) \geq 1 - \frac{\sigma^2}{\varepsilon^2}$$

Comment on Proposition 5.5: Let's examine what the proof looks like in the continuous case. (The discrete case may be proved similarly by replacing the integrals by sums.)

First notice that it is sufficient to prove the proposition for the special case in which $E(X) = 0$. The reason is that the random variables X and $X - \mu$ have the same variance. Therefore, if the proposition is true for $X - \mu$, it will be true for X also.

So now suppose that X is a continuous random variable with mean $\mu = 0$ and with density function f; then

$$\begin{aligned} P(|X| \geq \varepsilon) &= P(X \geq \varepsilon) + P(X \leq -\varepsilon) \\ &= \int_{-\infty}^{-\varepsilon} f(t)\, dt + \int_{\varepsilon}^{\infty} f(t)\, dt \\ &\leq \int_{-\infty}^{-\varepsilon} \frac{t^2}{\varepsilon^2} f(t)\, dt + \int_{\varepsilon}^{\infty} \frac{t^2}{\varepsilon^2} f(t)\, dt \\ &= \frac{1}{\varepsilon^2}\int_{-\infty}^{-\varepsilon} t^2 f(t)\, dt + \frac{1}{\varepsilon^2}\int_{\varepsilon}^{\infty} t^2 f(t)\, dt \end{aligned}$$

$$\leq \frac{1}{\varepsilon^2} E(X^2) = \frac{\sigma^2}{\varepsilon^2}$$

Example 5.5. A lot of 2,000 items has been produced on an assembly line on which 2% of the items produced are defective. Treating the assembly line as an independent trials process with each item produced having probability .02 of being defective, use Chebyshev's inequality to give a bound on the probability that in the batch of 2,000 items the number of defects is between 30 and 50.

Solution: Let X denote the number of defects in the 2,000 items. The assumption is that X is binomial with parameters $n = 2{,}000$ and $p = .02$. This enables us to compute μ and σ: $\mu = np = 40$ and $\sigma^2 = npq = 39.2$. If we now take $\varepsilon = 10$ in Chebyshev's inequality, we have

$$P(30 < X < 50) = P(|X - \mu| < 10) \geq 1 - \frac{39.2}{100} = .608$$

This should be viewed not so much as an estimate of the actual probability as a *bound* on the probability. The probability that X assumes a value within 10 of its mean is *at least* as great as .608. Since Chebyshev's inequality is based on nothing more than the mean and the variance of the random variable involved, it can't possibly give accurate estimates.

By way of comparison, we can relate this to the estimate provided by the central limit theorem (Proposition 5.4):

$$\begin{aligned} P(|X - \mu| < 10) &= P\left(\frac{|X - np|}{\sqrt{npq}} < \frac{10}{\sqrt{npq}}\right) \\ &\approx 2\Phi\left(\frac{10}{\sqrt{npq}}\right) - 1 \\ &\approx 2 \times .9441 - 1 \\ &\approx .888 \end{aligned}$$

As an illustration of the theoretical usefulness of Chebyshev's inequality, let's look further at what it has to say about an independent trials process. Denote by X_n the number of successes to occur during the first n trials of an independent trials process with success probability p. Chebyshev's inequality says then that

$$P(|X_n - np| < \varepsilon) \geq 1 - \frac{npq}{\varepsilon^2}$$

The value of this information is that it is true for whatever choice of $\varepsilon > 0$ that we wish to make. In particular, let's replace ε by $n\delta$, where δ is thought of as "small." Then

$$P\left(\left| \frac{X_n}{n} - p \right| < \delta \right) \geq 1 - \frac{npq}{n^2\delta^2} = 1 - \frac{pq}{n\delta^2}$$

Now notice that for any value of δ, the right side converges to 1 as $n \to \infty$. What does this mean? It means that for any choice of δ, *no matter how small*, the probability that the fraction of trials resulting in success differs from the theoretical probability p by less than δ tends to 1 as the number of trials increases without bound. In other words, it is guaranteed that the *observed* relative frequency of successes will converge to the *theoretical* relative frequency (as measured by p) as the number of trials tends to ∞. This is a special case of an important theoretical result called the *law of large numbers*.

5.4 Evaluating the Standard Normal Distribution Function

Since the antiderivatives of the standard normal density function cannot be expressed in terms of elementary functions, it is necessary to use means other than the fundamental theorem of calculus to evaluate the integrals that define values of the standard normal distribution function.

The relation between the standard normal distribution function Φ and the standard normal density function is given by

$$\Phi(s) = \int_{-\infty}^{s} f(t)\, dt$$

where f is the standard normal density function

$$f(t) = \frac{1}{\sqrt{2\pi}}\, e^{-t^2/2}$$

Thus, if values for the standard normal distribution function are needed, what is necessary is a good way to approximate this improper integral. In fact, the symmetry of the integrand $f(t)$ and the fact that f is a density function mean that

$$\int_{-\infty}^{\infty} f(t)\, dt = 1 \quad \text{and} \quad \int_{-\infty}^{0} f(t)\, dt = \frac{1}{2}$$

So if we want to compute $\Phi(s)$ for some $s > 0$, then

$$\Phi(s) = \frac{1}{2} + \int_0^s f(t)\,dt$$

On the other hand, if $s > 0$, then the symmetry of $f(t)$ about the y-axis implies that

$$\Phi(-s) = \int_{-\infty}^{-s} f(t)\,dt = \int_s^{\infty} f(t)\,dt = 1 - \int_{-\infty}^{s} f(t)\,dt = 1 - \Phi(s)$$

Thus all we really need is a way of evaluating $\int_0^s f(t)\,dt$ for $s > 0$.

Problems about normal random variables, of course, don't have to be phrased in terms of the normal distribution function at all. Rather, they can be expressed in terms of the normal density function. For example, if X is standard normal and we need to know $P(.5 < X < 1)$, we might just as well think of integrating (by some numerical method) the standard normal density from .5 to 1 rather than thinking of $P(.5 < X < 1)$ as $\Phi(1) - \Phi(.5)$.

In summary, all that is needed to get approximate values for normal random variables is some means of approximating definite integrals of the standard normal density function over finite intervals. It is not necessary to approximate improper integrals.

Simpson's Rule

Most calculus students learn *Simpson's rule* as a numerical method for approximating definite integrals. The idea behind Simpson's rule is very simple: Given a function defined on an interval, plot the value of the function at the two endpoints and at the midpoint and then draw the unique parabola through the three points. The area under the parabola is used to approximate the area under the curve. To obtain accuracy, the original interval may be split into smaller subintervals and this procedure used on the subintervals.

Crude as it sounds, this procedure works beautifully for smooth functions such as the standard normal density. For example, if we wish to approximate the probability $P(0 < X < 1)$ using this method, where X is standard normal, and if we use the two subintervals, [0, 1/2] and [1/2, 1], then the approximation obtained is .341355. The true value, correct to 6 digits, is .341344. The arithmetic of Simpson's rule in this particular scenario boils down to computing

$$\frac{1}{12} \times \{ f(0) + 4f(.25) + 2f(.5) + 4f(.75) + f(1) \}$$

where f is the standard normal density function. Such a calculation is easy with a scientific calculator and even easier with a programmable one. Students who know Simpson's rule are encouraged to try it out as a method of getting numerical answers in exercises involving normal random variables.

Other Available Numerical Methods

Since the mid 1970s many scientific calculators have included the normal distribution function as a built-in function. Some program must then reside within the calculator for computing the values. The one described in Problem 5.14 has been used in Texas Instruments™ calculators for a number of years. It is easily programmed on any programmable calculator or pocket computer.

Tables

Before calculators were available, it was common practice to consult tables that give values of the normal distribution function. This practice is rapidly becoming as outdated as the practice of looking up square roots in a table. Nevertheless, a table is included as Appendix B at the end of this book.

Problems

5.1 Show that if X is uniform on $[a, b]$, then

$$E(X) = \frac{a+b}{2} \quad \text{and} \quad \text{var}(X) = \frac{(b-a)^2}{12}$$

5.2 A mechanical device is known to have an expected lifetime of 10 years. Assume that the lifetime for such devices is known to have an exponential distribution.

(a) What is the probability that a device of this type will last more than 5 years?

(b) What is the probability that such a device will fail during the first 10 years?

(c) If such a device has already lasted 10 years, what is the probability that it will fail during the next 10 years?

5.3 Suppose we have a random number generator that generates numbers between 0 and 1 according to the uniform density on the interval [0, 1]. If we generate such a random number X and then consider $Y = 1 - X$, what kind of density function does Y have? (Your intuition might lead you to guess the answer before doing the calculation.)

5.4 Suppose you have just bought a piece of equipment that has been advertised to have an expected lifetime of 2 years. Is it likely to actually last that long? (In other words, how probable is it that the device will actually last that long?)

(a) Assume that the lifetime is normally distributed.

(b) Assume that the lifetime is exponentially distributed.

5.5 Assume that X has an exponential density with parameter $\lambda = 1$.

(a) Find F_X.

(b) If $Y = 2X + 1$, find F_Y.

(c) Find the density for Y.

(d) Using Definition 3.7, compute $E(Y)$.

(e) Compute $E(Y)$ again, this time using Proposition 3.7.

5.6 Suppose X is a uniform random variable on [0, 1]. Let $Y = X^2$. Compute the distribution function for Y and the density function for Y. Notice that f_Y is an unbounded function, but that part 2 of Definition 3.3 is satisfied by f_Y if the integral is interpreted as an improper integral. Compute $E(Y)$ using Definition 3.7 and again using Proposition 3.7.

5.7 Assume X is uniform on [0, 2], but now suppose $Y = |X - 1|$. Determine the distribution function, the density function, and the expected value of Y. [Hint: The biggest difference between this and Problem 5.6 is that you will have to be more careful and clever in trying to write F_Y in terms of F_X.] One interpretation of this problem is to think of it as a model for choosing a random number between 0 and 2 and then to consider how close the number chosen is to 1.

5.8 A wire is 2 ft long. A random point on the wire is picked, and the wire is cut at that point.

(a) What is the expected value for the length of the shortest piece?

(b) If Y is the random variable that denotes the length of the shortest piece of wire obtained when the original is cut, find the distribution function and the density function for Y.

[Hint: This problem is closely related to Problem 5.7. Think of the wire as lying on the interval [0, 2] and X as being the point where the cut is made. Assume X is uniform on [0, 2]. Now write Y as a function of X.]

5.9 A device has a lifetime which is known to be an exponential random variable X with $E(X) = 10$ years. Find the value of t_0 for which the probability is exactly 1/2 that the device lasts at least t_0 years; that is, $P(X > t_0) = 1/2$.

5.10 Suppose X is an exponential random variable with parameter $\lambda = 1$. Determine the distribution function for the random variable $Y = X^2$.

5.11 If X is normal with parameters $\mu = 5$ and $\sigma = 2$, find $P(2 < X < 4)$. You can use Simpson's rule, the numerical method of Problem 5.14, or the table in Appendix B.

5.12 Show that if X is an exponential random variable with parameter λ, then the variance of X is $1/\lambda^2$.

5.13 Use the normal approximation to the binomial (Proposition 5.4) to approximate the probability of between 290 and 305 successes in 1,000 trials of an independent trials process with $p = .3$.

5.14 It used to be commonplace to consult long and tedious tables when values of the standard normal distribution function were needed. Many scientific calculators now have this function built in. In addition, there are many common and simple numerical methods that may be used to evaluate it. The following peculiar but useful formula is given in the *Handbook of Mathematical Functions* of the National Bureau of Standards.

If Φ denotes the standard normal distribution function and

$$f(t) = \frac{1}{\sqrt{2\pi}}\, e^{-t^2/2}$$

denotes the standard normal density function, then values of Φ may be approximated via the numerical approximation

$$\Phi(t) \approx 1 - f(t)\ (a_1\beta + a_2\beta^2 + a_3\beta^3 + a_4\beta^4 + a_5\beta^5)$$

where $\beta = 1/(1 + .2316419\ t)$ and where $a_1 = .31938153$, $a_2 = -.356563782$, $a_3 = 1.781477937$, $a_4 = -1.821255978$, and $a_5 = 1.330274429$. This approximation is easy to program (see Appendix B) and quite accurate when $t > 0$. Try it out with a scientific calculator in Example 5.2, where $\Phi(-.5)$ is needed. Compute $\Phi(.5)$ using the method described here. Then $\Phi(-.5) = 1 - \Phi(.5)$.

5.15 Assume that the number of defects in a batch of 10,000 manufactured items is a normal random variable with mean $\mu = 300$ and standard deviation $\sigma = 100$. (Clearly the actual number is an integer-valued random variable. By now you should be getting used to the idea of using a continuous model for a situation that technically is discrete.) Find the probability that the actual number of defects found in a batch will be between 225 and 275.

5.16 If 100 items are tested from a batch in which 20% are bad, what is the probability that the number of bad ones found will be between 15 and 25 inclusive, that is, $P(15 \le X \le 25)$, where X = number of defectives found?

(a) If each one tested is considered an independent trial with probability $p = .2$ that the item is bad, then the number of bad items found is binomial with $n = 100$ and $p = .2$. Compute $P(15 \le X \le 25)$ assuming that X has this probability distribution.

(b) Now approximate this probability using the fact that $(X - np)/\sqrt{npq}$ is approximately standard normal, and compare your answer to the exact answer in part a. (If you are really interested in great accuracy in your approximation here, you might want to consider further the fact that X is actually integer-valued. This means that, in fact,

$$P(15 \le X \le 25) = P(14 < X < 26)$$

If one "averages" these two intervals and uses

$$P(14.5 < X < 25.5)$$

in calculating the normal approximation, then the approximation of the discrete binomial with the continuous normal will be very good.)

5.17 Suppose that a flagpole is erected and that it is assumed that the "lifetime" of the flagpole is a random variable having an exponential density function and

having an expected value of 50 years. Find
(a) the probability that the flagpole lasts more than 25 years,
(b) the probability that it lasts more than 50 years, and
(c) the probability that it lasts more than 100 years.

5.18 We will approach Problem 5.17 from another viewpoint. Instead of measuring time "continuously," we will measure it "discretely." Suppose that each year there is a probability .02 that the flagpole will fall during that year, and suppose that this is viewed as an "independent trials process." Let X denote the number of the year in which the flagpole finally falls, so X could assume values 1, 2, 3, $\cdots$ (X is a discrete random variable.) If we consider the years to be independent trials, then X can be viewed as a geometric random variable with $p = .02$. Under these assumptions, compute
(a) the probability that the flagpole lasts at least 25 years,
(b) the probability that it lasts at least 50 years, and
(c) the probability that it lasts at least 100 years.

5.19 Prove Proposition 5.2. [Hint: The mean is easy. To compute the variance, use integration by parts with $u = t$ and $dv = t\,e^{-t^2/2}$. Then to evaluate

$$\int v\,du$$

in the integration-by-parts formula, where

$$\int u\,dv = uv - \int v\,du$$

use the fact that the standard normal density *is* a density function.]

5.20 Prove Proposition 5.3. [Hint: In part 1 do a change of variables to convert to the standard normal density. In part 2 show that the distribution function of $(X - \mu)/\sigma$ is the standard normal distribution function. In part 3 a similar change of variables is useful.]

5.21 Suppose that the number of cases of a certain disease per 100,000 population is a normally distributed random variable with mean 300 and standard deviation 100. What is the probability that a particular community of 100,000 people would have more than 500 cases of the disease? (Notice that this is another instance where we are using a continuous model for what is actually a discrete phenomenon.)

5.22 Suppose a random number is generated (uniform on [0, 1]) and then truncated after the first digit so the result is one of the numbers 0, .1, .2, · · · , .9. Compute the expected value for this truncated result, and understand your computation in the context of Proposition 3.7. One interesting feature of this situation is that we are looking at an instance of the relation $Y = g(X)$, where X is continuous and Y is discrete.

5.23 Suppose that $\alpha > 0$ and that X is a random variable that is uniformly distributed on the interval [0, 1]. Find the distribution function and the density function for the random variable

$$Y = -\frac{1}{\alpha} \ln (1 - X)$$

(This problem has very useful applications, because many devices such as calculators and computer routines are equipped with random number generators that produce numbers between 0 and 1 according to a uniform density on [0, 1]. If one wants a "random" value for an exponentially distributed random variable and has such a "uniform on [0, 1]" random number generator available, this exercise shows how to get one.)

5.24 A random number between 0 and 1 is produced using the uniform density on [0, 1]. What is the probability that the first digit in the decimal representation of the square root of the number is a 7?

5.25 In the circuit pictured below, the resistance of the resistor labeled R fluctuates between 1 and 2 ohms. Assume a uniform distribution for the resistance between 1 and 2 ohms. If the battery output is a constant 12 volts, what is the probability distribution of the voltage drop across this resistor? Find the distribution function and the density function for this voltage drop. What is the expected value for this voltage drop across R?

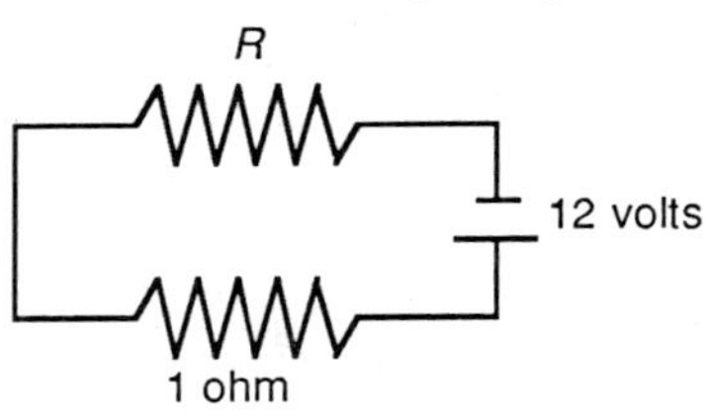

5.26 Resistor R_1 in the following figure has a constant resistance of 1 ohm. Resistor R_2 has resistance which varies uniformly over the range of 1 to 2 ohms. Find the density function for the resistance in the circuit, that is, the resistance of the two resistors wired in parallel.

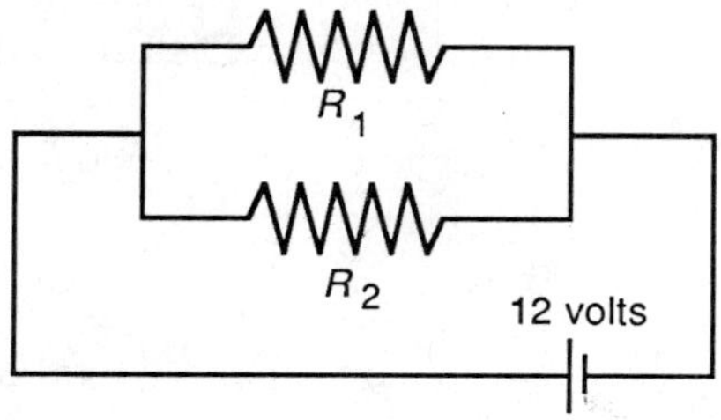

5.27 (a) A random variable X gives the voltage output from an acoustical transducer. The voltage varies from –3.5 to +3.5 volts. Assume that the probability distribution for X is the uniform distribution on the interval [–3.5, 3.5]. Sketch a graph of the cumulative distribution function for X.

(b) Suppose now that for some reason it is desirable to round off the voltage to the nearest integer. In this case, the voltage will be between –3 and +3 volts. This integer-valued voltage random variable can be written as $Y = \text{round}(X)$, where *round* is the round-off function; that is, $\text{round}(x) =$ the value of x rounded off to the nearest integer for any real number x. Determine the probability mass function for this discrete random variable Y, and then observe that this is the same probability distribution as the one encountered in Problem 4.12.

5.28 A manufacturer wishes to estimate the proportion of defects in a batch of items. If n items are to be tested and X is the number of defects found, then X/n will be used as the estimate of p, the fraction of the lot that consists of defective items. Using techniques parallel to those of Example 5.4, determine how large n must be, that is, how many must be tested in order to be 90% certain that the observed value of X/n is actually within .05 of the true value of p.

5.29 Redo the calculation in Example 5.3 using the continuity correction (as described in that example) to see how much it changes the answer.

5.30 Show that if Z is a standard normal random variable, then $P(|Z| > d) = 2 - 2\Phi(d)$ for any number $d > 0$.

Supplemental Exercises on the Normal Distribution

5.31 If X is a standard normal random variable, find $P(X < 1.5)$.

5.32 If X is standard normal, find $P(X > .22)$.

5.33 If X is normal with mean $\mu = 15$ and standard deviation $\sigma = 3$, find the probability $P(X < 17)$.

5.34 If X is a normal random variable, find the probability that the value assumed by X lies more than 2 standard deviations away from the mean. [Hint: Part of the problem here is to convince yourself that the answer to this question is the same regardless of what the mean and variance of X happen to be.]

5.35 If X is standard normal, find (approximately) the value of t for which it is true that $P(X > t) = .95$. [Hint: If you look at this problem carefully, you should see that this question is equivalent to the problem of evaluating the *inverse* of the standard normal distribution function at .05. You can either use a table, or if you have a routine for calculating values of the standard normal distribution, you can by trial and error zero in on an approximate answer to this question.]

5.36 If X is normal with mean $\mu = 3$ and standard deviation $\sigma = 1$, find a 95% confidence interval for X centered at 3. This is a fancy way of saying the following: Find the number $\delta > 0$ having the property that

$$P(3 - \delta < X < 3 + \delta) = .95$$

(See the hint in Problem 5.35. This question can also be phrased in terms of finding values of the inverse of the standard normal distribution function.)

Chapter 6: Joint Distributions

Many common mathematical models involve more than one quantity. For example, a model for a gas is concerned with the interaction of temperature and pressure. When dealing with a probabilistic model, this means that we must be concerned with the way that two or more random variables interact. Knowledge of the individual behavior of the two will not be adequate because we may need to know how certain values assumed by one affect the probability of the other assuming given values. In the discrete case the interaction is studied via the joint probability mass function, and in the continuous case via the joint density function. There is additionally a natural way to extend the concept of the distribution function to this "joint" setting, but the joint distribution function is less useful for computation and will not be emphasized in this book.

When studying the relation between two random variables, we often will need to consider the intersection of two events of the form $(X = a)$ and $(Y = b)$. A useful way to denote the intersection of these two events is to write the intersection as $(X = a, Y = b)$. The probability of this event is written $P(X = a, Y = b)$. Expressions such as $P(a < X < b, c < Y < d)$ have a similar meaning, that is, the event whose probability is being represented is the *intersection* of the two events separated by commas.

6.1 Joint Probability Mass Functions

Definition 6.1: If X and Y are discrete random variables, then their *joint probability mass function* $p_{X,Y}$ is the function of two variables defined by

$$p_{X,Y}(x, y) = P(X = x, Y = y)$$

for each pair of real numbers x and y.

Example 6.1. Four items are labeled with the numbers 0, 1, 2, and 3, and two of the items are randomly selected one at a time without replacement. We will denote the number on the first item selected by X, and the number on the second by Y.

It is sometimes useful, in examples such as this where there are a small number of possible outcomes, to exhibit all values of the joint mass function in a table such as the one shown in Figure 6.1, which presents everything there is to say about the joint mass function for X and Y.

	0	1	2	3	***X***
0	0	1/12	1/12	1/12	**1/4**
1	1/12	0	1/12	1/12	**1/4**
2	1/12	1/12	0	1/12	**1/4**
3	1/12	1/12	1/12	0	**1/4**
Y	**1/4**	**1/4**	**1/4**	**1/4**	

Figure 6.1 Joint probability mass function for the random variables X and Y of Example 6.1.

The 16 numbers in the table that are not in boldface give all values of $p_{X,Y}$. For example the first 1/12 in the top row gives the probability $P(X = 1, Y = 0)$. The boldface numbers at the bottom and at the right edge are the values for the mass function of X (bottom row) and Y (rightmost column). Notice that these probabilities may be obtained simply by adding the values of the joint mass function in the corresponding row or column. The reason is elementary. For example, the event $(X = 1)$ is the disjoint union

$$(X = 1, Y = 0) \cup (X = 1, Y = 1) \cup (X = 1, Y = 2) \cup (X = 1, Y = 3)$$

and therefore,

$$P(X = 1) = P(X = 1, Y = 0) + P(X = 1, Y = 1) + P(X = 1, Y = 2) + P(X = 1, Y = 3)$$

Proposition 6.1 summarizes the relation between p_X, p_Y, and $p_{X,Y}$. The proof of the proposition is based on the same ideas that are illustrated in the example above. Conceptually, it is important to understand that the joint probability mass function carries all the information necessary to do computations related to the interaction of X and Y. In particular, Proposition 6.1 summarizes how the mass functions of X and Y individually may be determined from the joint mass function. Often the probability mass functions of X and Y are referred to as the *marginal* probability mass functions of X and Y to distinguish them from the *joint* mass function for the pair X, Y.

Proposition 6.1: The relationship between the marginal mass functions p_X and p_Y and the joint mass function $p_{X,Y}$ can be described in general as follows:

1. For any number x, $p_X(x) = \sum p_{X,Y}(x, y_k)$ where the sum extends over all values y_k assumed by Y.
2. For any number y, $p_Y(y) = \sum p_{X,Y}(x_k, y)$ where the sum extends over all values x_k assumed by X.

We have encountered the idea of independence of events in Chapter 1. It is also useful to have an understanding of what independence means in the context of random variables.

Definition 6.2: Two discrete random variables X and Y are *independent* if it is true that

$$P(X = x, Y = y) = P(X = x)\,P(Y = y)$$

for all choices of x and y.

The basic idea is that two random variables are independent if events described in terms of one random variable are independent from events described in terms of the other. Literally, the definition says that X and Y will be considered independent random variables provided that events of the form $(X = x)$ and $(Y = y)$ are independent events for all choices of numbers x and y.

Example 6.2. One 4-ohm resistor and two 8-ohm resistors are in a box. A resistor is randomly drawn from the box and inspected; then it is replaced in the box and a second is drawn. We will denote by X the resistance of the first one drawn from the box and by Y the resistance of the second. Since this is sampling *with* replacement, the result of one test has no influence on the result of the other test. This is indicated in the following table of values of the joint mass function for X and Y. In this case X and Y are independent random variables.

When X and Y are independent, the values of the joint mass function are determined by the values of the mass functions of X and Y. For this example, this is clearly visible in Figure 6.2. The circled entries illustrate that $2/9 = 1/3 \times 2/3$. Similarly, each of the four values of the joint mass function shown in Figure 6.2 is the product of the appropriate value of the mass function of X and the mass function of Y. In general (when independence is lacking), it is not possible to construct the joint probability mass function from a knowledge of the mass functions of X and Y individually. See Problem 6.31(a) for an interesting illustration of this.

	4	8	X
4	1/9	2/9	1/3
8	2/9	4/9	2/3
Y	1/3	2/3	

Figure 6.2 Joint mass function for the independent random variables of Example 6.2. The circled entries illustrate that $P(X = 8, Y = 4) = P(X = 8)\, P(Y = 4)$.

6.2 Joint Density Functions

The joint density for a pair of continuous random variables will also necessarily be a function of two variables. It plays the same role that the joint mass function plays for a pair of discrete random variables. That the probability of the entire sample space be 1 requires the double integral over the xy-plane be 1, just as the integral in Definition 3.3 in Chapter 3 is required to be 1.

Definition 6.3: A function f of two variables is said to be a (two-dimensional) *density function* provided that

1. $f(x, y) \geq 0$ for all x and y, and

2. $\int_{-\infty}^{\infty}\int_{-\infty}^{\infty} f(x, y)\ dx\ dy = 1$

In order to motivate the relation between a pair of continuous random variables and their joint density function, it is worth observing that in the discrete case the probability $P(a \leq X \leq b,\ c \leq Y \leq d)$ is, in fact, equal to the sum

$$\sum p_{X,Y}(x_j, y_k)$$

where the sum here extends over all pairs (x_j, y_k) where x_j and y_k satisfy the conditions $a \leq x_j \leq b$ and $c \leq y_k \leq d$. In fact, if B is any region in the plane and if we used the notation $(X, Y) \in B$ as a short way to denote the event $\{w \in S : (X(w), Y(w)) \in B\}$ in the underlying sample space S, then we can write the probability of this event as

$$P\{(X, Y) \in B\} = \sum_{(x_j, y_k) \in B} p_{X,Y}(x_j, y_k) \tag{6.1}$$

Problem 6.10 is a simple exercise which will check your understanding of what Equation 6.1 says.

Perhaps now you can guess the definition of the joint density function for a pair of continuous random variables by simply thinking of summation in the discrete case being replaced by integration in the continuous case.

Definition 6.4: Given a pair of random variables X and Y, to say that a function f (of two variables) is the *joint density function* for X and Y means that

1. the function f is a density function (see Definition 6.3), and

2. $P\{(X,Y) \in B\} = \iint_B f(x,y)\ dx\ dy$ whenever B is a region in the xy-plane.

Among the easiest to understand examples of joint density functions are the joint uniform densities.

Definition 6.5: Given a region T in the xy-plane, the *uniform joint density function* on the region T is the function f defined by

$$f(x, y) = \begin{cases} \frac{1}{A} & \text{if } (x,y) \in T \\ 0 & \text{if } (x,y) \notin T \end{cases}$$

where A denotes the area of the region T.

The reason that the constant value that f assumes on T must be the reciprocal of the area of T is that the integral of the constant function 1 over the region T is precisely the area of T. So this requirement makes the joint uniform density satisfy condition 2 of Definition 6.3.

Example 6.3. Suppose X and Y have as joint density function the uniform density on the square $T = \{(x, y) : 0 \leq x \leq 1, 0 \leq y \leq 1\}$. What is the probability that X takes on a value larger than Y? In other words, what is the probability of the event $(X > Y)$?

Solution: The main step that is necessary to grasp this example is to understand what the event $(X > Y)$ means. It means the set of all elements w in whatever sample space X and Y are defined on which have the property that the pair of number $X(w)$ and $Y(w)$ satisfy the condition $X(w) > Y(w)$. In order to relate this to Definition 6.4, it is necessary to rephrase this as follows: We are considering the event that consists of all elements of the sample space for which $(X(w), Y(w))$, when viewed as a point in the xy plane, lies to the right of the line $y = x$. So we are interested in $P\{(X, Y) \in B\}$, where B is the half-plane that is shaded in the following picture.

From Definition 6.4, what is needed in order to compute $P(X > Y)$ is to integrate the joint density over the region B. In this example, however, since the joint density is 0 off the square T, the integral over the region B can be reduced to the integral over $B \cap T$.

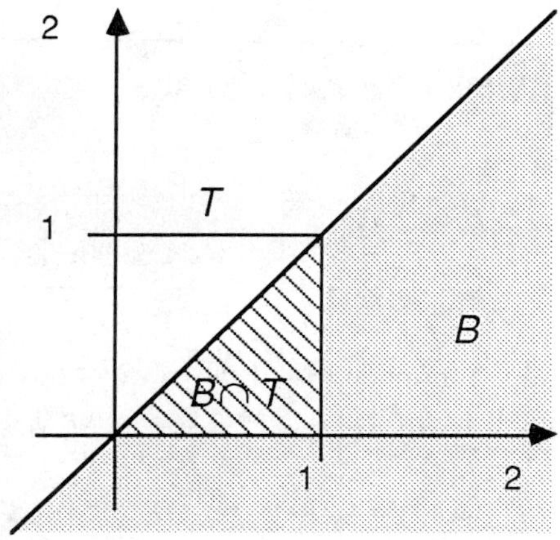

This train of thought is summarized in the following steps:

$$
\begin{aligned}
P\{(X, Y)\in B\} &= \iint_B f(x, y)\ dx\, dy \\
&= \iint_{B\cap T} f(x, y)\ dx\, dy \\
&= \iint_{B\cap T} 1\ dx\, dy \\
&= \frac{1}{2}
\end{aligned}
$$

(Since we are integrating the constant function 1 over the triangle $B \cap T$, the value of the integral is just the area of the triangle.)

Recall how, in the discrete case, the mass functions for X and Y can be constructed by adding the appropriate values of the joint mass function. A similar method is available in the continuous case for obtaining the density functions for X and Y from the joint density. This method is described in Proposition 6.2, which is similar to Proposition 6.1 except that (as usual) summation is replaced by integration. Often the density functions of X and Y are referred to as the *marginal densities* to distinguish them from the joint density for X and Y.

Proposition 6.2: If X and Y have joint density function f, then the densities for X and Y may be obtained from f as follows:

$$f_X(x) = \int_{-\infty}^{\infty} f(x, y)\, dy \quad \text{and} \quad f_Y(y) = \int_{-\infty}^{\infty} f(x, y)\, dx$$

Proof: Given any numbers a and b with $a < b$, $P(a < X < b)$ can be computed from the joint density function using Definition 6.4 by taking B to be the region

$$\{(x, y) : a < x < b, -\infty < y < \infty\}$$

that is, B is an infinite vertical strip in which no restriction is placed on the second coordinate. Then

$$P(a < X < b) = P\{(X, Y) \in B\} = \iint_B f(x, y)\, dx\, dy$$

$$= \int_a^b \int_{-\infty}^{\infty} f(x, y)\, dy\, dx = \int_a^b g(x)\, dx$$

where $g(x)$ is the function

$$g(x) = \int_{-\infty}^{\infty} f(x, y)\, dy$$

But knowing that

$$P(a < X < b) = \int_a^b g(x)\, dx$$

for all choices of a and b with $a < b$ is exactly what we need to know in order to conclude that g is the density function for X. (See Definition 3.4.)

Clearly the proof of the second statement involving the density of Y can be given in a similar fashion.

Example 6.4. In a certain device, X represents the voltage drop across a component and Y represents the current in another part of the device. The two are related probabilistically, however. Let's suppose that X and Y have a joint density function $f(x, y) = c$ in the part of the first quadrant bounded by the curve $y = 4 - x^2$ and the x axis, with $f(x, y) = 0$ outside this region.

(a) Find the value c must have in order to make this a joint density.
(b) Find the marginal densities f_X and f_Y.
(c) What is the probability that $3X$ is greater than Y?

Solution: (a) $$\int_{-\infty}^{\infty}\int_{-\infty}^{\infty} f(x, y)\, dx\, dy = \int_0^2\int_0^{4-x^2} c\ dy\, dx = 1$$

Carry out this computation and solve for c, obtaining $c = 3/16$.

(b) If $0 < x < 2$ then

$$f_X(x) = \int_{-\infty}^{\infty} f(x, y)\, dy = \int_0^{4-x^2} \frac{3}{16}\, dy = \frac{3(4-x^2)}{16}$$

Furthermore, if x is not between 0 and 2, then $f_X(x) = 0$. If $0 < y < 4$, then

$$f_Y(y) = \int_{-\infty}^{\infty} f(x, y)\, dx = \int_0^{\sqrt{4-y}} \frac{3}{16}\, dx = \frac{3\sqrt{4-y}}{16}$$

Moreover, $f_Y(y) = 0$ if y is not between 0 and 4.

(c) $P(Y < 3X)$ is computed by integrating the joint density function over the shaded region in the picture that follows.

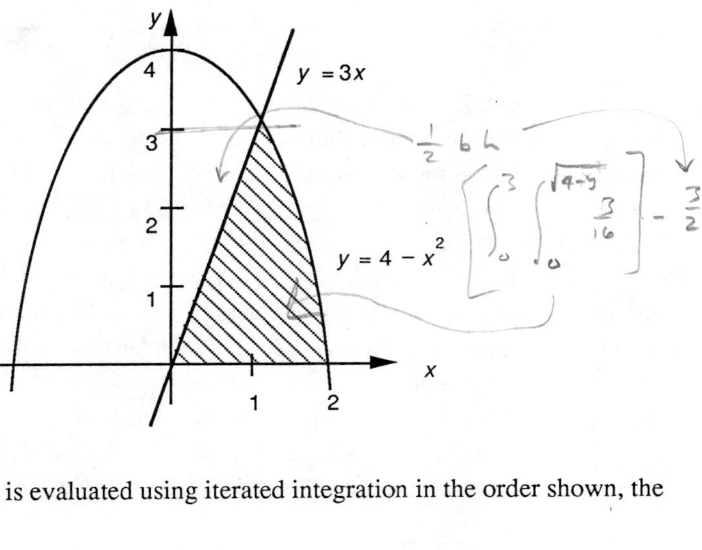

If the double integral is evaluated using iterated integration in the order shown, the endpoints of integration for the integration with respect to y are 0 and 3. This reason is that the y coordinate of the point where the line $y = 3x$ and the parabola $y = 4 - x^2$ intersect is 3.

$$P(Y < 3X) = \int_0^3 \int_{y/3}^{\sqrt{4-y}} c \; dx \, dy$$

$$= \int_0^3 \sqrt{4-y} - \frac{y}{3} \; dy = \frac{19}{32}$$

What should be meant by independence of continuous random variables? The most natural guess might be to require something like

$$P(a < X < b \, , c < Y < d) = P(a < X < b) \, P(c < Y < d)$$

for all values of a,b,c, and d. This approach would be satisfactory (see Problem 6.11), but it is a bit easier to take the following definition, which is essentially equivalent and is easier to use.

Definition 6.6: Two continuous random variables X and Y are *independent* provided that the function f defined by

$$f(x, y) = f_X(x) \, f_Y(y)$$

is a joint density function for X and Y.

Example 6.5. Let's reconsider X and Y in Example 6.3. With Proposition 6.2 it is easy to check that both X and Y have as their density functions the uniform density on the interval [0, 1]. Furthermore, it is quite apparent now that the joint density f does satisfy $f(x, y) = f_X(x) \, f_Y(y)$, for both sides are equal to 1 when $0 \leq x \leq 1$ and $0 \leq y \leq 1$, and both sides are equal to 0 otherwise.

This example is a model for the following experiment: Suppose you have a random number generator that will generate two random numbers between 0 and 1 in such a way that the numbers produced are independent of each other and so that the probability distribution for each is the uniform distribution on the interval [0, 1]. A variety of interesting questions can be asked about the two numbers obtained as output from the random number generator. Problems 6.8 and 6.12 are two such questions, and a further application of the same model is found in Example 6.9.

Example 6.6. Two devices manufactured by different companies are known to have exponentially distributed lifetimes. One has an expected lifetime of 5 years and the other 8 years. What is the probability that the one with the longer expected lifetime will actually last more than twice as long as the other? (Assume that the two devices function independently.)

Solution: Let X be the lifetime of the device with the expected lifetime of 5 years and Y the lifetime of the device with expected lifetime of 8 years. This simply says $E(X) = 5$ and $E(Y) = 8$. Furthermore, since both have exponential distributions, we know from Problem 3.15 that X is exponential with parameter $\lambda = .2$ and Y is exponential with parameter $\lambda = .125$.

Since X and Y are to be assumed independent, we know that the function $f(x, y) = f_X(x)\, f_Y(y)$ will be a joint density function for the pair. The question posed is what is the probability that $Y > 2X$. The picture guides the calculation.

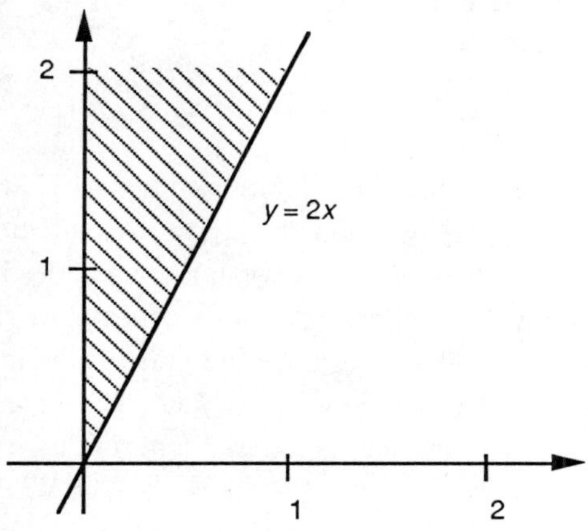

The event $(Y > 2X)$ can be viewed as $\{(X, Y) \in B\}$ where B is the region consisting of the half-plane above the line $y = 2x$ in the picture. So the probability of this event can be computed by integrating the joint density over this region. However, the joint density function itself is equal to 0 except in the first quadrant. Therefore the integral to be computed actually can be reduced to an integral over the wedge shaped slice of the first quadrant shown in the picture. The computation then goes as follows:

$$P\{(X, Y) \in B\} = \iint_{wedge} f(x, y)\, dx\, dy$$

$$= \int_0^\infty \int_{2x}^\infty (.2\, e^{-.2x})\, (.125\, e^{-.125y})\; dy\, dx$$

Problem 6.16 asks you to complete this calculation.

There is not a great conceptual hurdle in going from independence of two random variables to independence of more than two. It is simply a matter of passing from two dimensions to n dimensions, where n is the number of random variables being considered. For example with continuous random variables one first has to extend Definition 6.3 so as to define an n-dimensional density function with the analogous properties. (The integral in that case becomes an integral over n-dimensional space.) Definition 6.4 could then be extended to define the joint density of random variables $X_1, \cdots, X_n$, and the integral in that definition would pertain to regions B in n-dimensional space. Finally, an extended form of Definition 6.6 would say that $X_1, ..., X_n$ are independent provided that their joint density function is given by

$$f(x_1, \cdots, x_n) = f_{X_1}(x_1) \cdots f_{X_n}(x_n)$$

In this book we will not do computations involving such n-dimensional density functions. In the next section, however, we will be examining some situations where more than two independent random variables interact. While the technical definitions involve concepts such as those illustrated in the above paragraph, the most important idea you need to grasp is the intuitive meaning of independence of more than two random variables. The idea is that knowledge about values assumed by certain ones is independent (in the sense of independent events) from knowledge about others. For example, if X_1, X_2, X_3, and X_4 are independent, this means, for instance, that the values assumed by X_1 and X_3 would have no effect on the values assumed by X_2 and X_4.

The reason this intuitive understanding is important is that in constructing a mathematical model, one decision that often must be made is the decision as to whether certain random variables can be considered to be independent. Historically, some of the most serious errors in constructing complex probability models (for example, safety analyses of complicated systems such as aircraft or nuclear reactors) have come about because certain random variables were assumed to be independent when, in fact, they were not. Section 6.4 will introduce various scenarios where the concept of independence is important and useful.

6.3 Functions of Two Random Variables

Just as we found it helpful in Chapter 3 to consider composite random variables of the form $Y = g(X)$, it is now useful to consider composite random variables of the form $Z = g(X, Y)$, where g is a function of two real variables. Such random variables have occurred earlier in examples. For instance, if two dice are rolled and if X and Y represent the numbers on the respective dice and Z the sum, then clearly $Z = X + Y$. This simply says $Z = g(X, Y)$ where g is the function of two variables $g(x, y) = x + y$.

If $Z = g(X, Y)$, it is useful to be able to compute the expected value of Z directly from knowledge of the function g and the joint distribution of X and Y. Propositions 3.6 and 3.7 have equally powerful versions for functions of two random variables provided that the two are discrete or, in the continuous case, that they have a joint density function. While sometimes it is possible to determine the probability distribution of Z from a knowledge of g, X, and Y, Proposition 6.3 offers an alternative that is usually easier if it is only the expected value that is needed.

Proposition 6.3:

1. If X and Y are discrete random variables and if $Z = g(X, Y)$, then the expected value of Z is given by

$$E(Z) = \sum_{x_j,\, y_k} g(x_j, y_k)\, p_{X,Y}(x_j, y_k)$$

where the sum extends over all ordered pairs x_j, y_k where x_j is a value assumed by X and y_k is a value assumed by Y.

2. If X and Y are continuous random variables having joint density function f and if $Z = g(X, Y)$, then

$$E(Z) = \int_{-\infty}^{\infty} \int_{-\infty}^{\infty} g(x, y)\, f(x, y)\; dx\, dy$$

Example 6.7. Two signals will occur at a random time during a one-hour period. Assume that they are timed independently and that the time of occurrence of each is uniformly distributed throughout the one-hour interval. What is the expected length of time between the occurrences of the two signals?

Solution: Let X and Y denote the times when the two signals occur. The assumption is that both X and Y are uniform random variables on [0, 1] and that they are independent. Their joint density then is the uniform density on the unit square, as in Example 6.3. The length of time between occurrence of the two signals is the random variable $Z = |X - Y| = g(X, Y)$ where g is the function $g(x, y) = |x - y|$. So the expected value of Z can be computed by performing the integration

$$E(Z) = \int_{-\infty}^{\infty}\int_{-\infty}^{\infty} |x - y|\ f(x, y)\ dx\,dy = \int_0^1\int_0^1 |x - y|\ dx\,dy$$

The trick to evaluating the integral on the right is to get rid of the absolute value signs by splitting the integral into two parts. The line $x = y$ separates the square into two triangles. Integrate over the unit square by integrating over the two triangles. This enables the absolute value signs to be removed, for on one of the triangles $x > y$ and so $|x - y| = x - y$, and on the other triangle $x < y$ and therefore $|x - y| = y - x$. When you carry out these details, you conclude that $E(Z) = 1/3$. (See Problem 6.18.)

Example 6.8 (Redundant safety systems). The picture below shows a pair of redundant safety switches that are installed to cut off the power to a device if a dangerous situation should quickly develop.

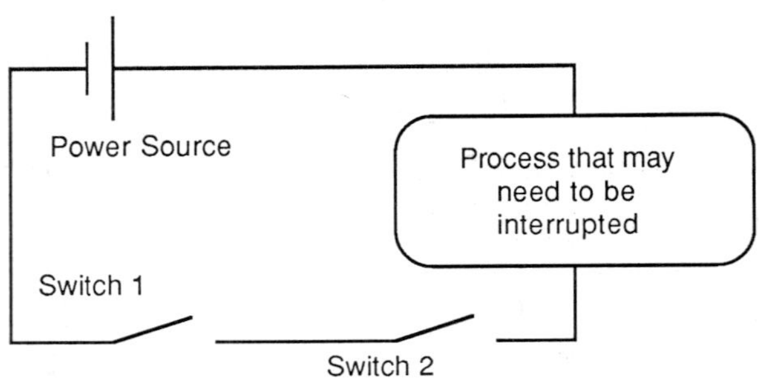

Different operators control the two switches, and so the two switches are viewed as functioning independently. There is a brief time period between the time that the dangerous situation arises and the time the switch is actually opened. We will assume that for both switches the delay time is uniformly distributed over a 5 second time interval. Clearly, if there were only one safety switch, the expected value for the length of time that elapses from the time that danger develops until the switch is actually thrown is 2.5 seconds. How much safety is actually achieved via the use of a backup operator controlling a second switch? In other words, using the two-switch system, what is the expected value for the elapsed time before a switch is opened to shut down the system?

Solution: We can let X denote the waiting time until switch 1 is opened and Y the waiting time until switch 2 is opened. The assumption is that X and Y are each uniformly distributed on the time interval [0, 5] and that they are independent. If Z is the waiting time until the current is actually cut off to the device, then $Z = g(X, Y)$, where g is the function of two variables

$$g(x, y) = \min\{x, y\}$$

that is, g is the function that selects the minimum of the two numbers input.

The relation between Z, X, and Y is that $Z = \min\{X, Y\}$. For any number s,

$$P(Z > s) = P(X > s, Y > s) = P(X > s)\, P(Y > s)$$

the last expression coming from the independence of X and Y. (See Problem 6.11.) In particular, if $0 < s < 5$, this probability is

$$P(Z > s) = \frac{5-s}{5} \times \frac{5-s}{5}$$

(Check this detail. It depends only upon understanding what the uniform distribution on the interval [0,5] looks like.)

Therefore,

$$F_Z(s) = P(Z \le s) = 1 - P(Z > s) = 1 - \frac{(5-s)^2}{25}$$

whenever $0 < s < 5$. The density function f_Z for Z may be obtained by differentiating F_Z. The two functions are as follows:

$$F_Z(s) = \begin{cases} 0 & \text{if } s < 0 \\ 1 - \dfrac{(5-s)^2}{25} & \text{if } 0 \le s \le 5 \\ 1 & \text{if } s > 5 \end{cases}$$

$$f_Z(s) = \begin{cases} \dfrac{10}{25} - \dfrac{2s}{25} & \text{if } 0 < s < 5 \\ 0 & \text{otherwise} \end{cases}$$

Once the density function for Z is known, $E(Z)$ can be calculated:

$$\begin{aligned} E(Z) &= \int_{-\infty}^{\infty} t\, f_Z(t)\, dt \\ &= \int_0^5 \frac{t\,(10-2t)}{25}\, dt \\ &= .04 \int_0^5 10t - 2t^2\, dt = .04\left(5t^2 - \tfrac{2}{3}t^3\right)\Big|_0^5 \\ &= \frac{5}{3} \end{aligned}$$

So the addition of the backup safety switch controlled by an independent operator reduces the expected time until shutdown from 2.5 seconds to 1.67 seconds.

Of course, we can also compute $E(Z)$ from Proposition 6.3, and this will be the easiest way to get the expected value if we have no particular interest in the probability density. Since $Z = \min\{X, Y\}$,

$$\begin{aligned} E(Z) &= \int_{-\infty}^{\infty}\int_{-\infty}^{\infty} g(x, y)\, f_{X,Y}(x, y)\ dx\, dy \\ &= .04 \int_0^5 \int_0^5 \min\{x, y\}\ dx\, dy \end{aligned}$$

The factor .04 in the last integral is the product of $f_X(x)$ and $f_Y(y)$, each of which is 1/5. The way to evaluate the integral on the right is to realize that the value of the function $g(x, y) = \min\{x, y\}$ is equal to y if (x, y) is a point lying below the line $y = x$ and is equal to x if (x, y) lies above the line $y = x$. So we split the double integral into two pieces as in the equation below, and then we compute

$$E(Z) = \int_0^5 \int_0^x .04y\ dy\, dx \quad + \quad \int_0^5 \int_x^5 .04x\ dy\, dx$$

These integrals are easily evaluated. Each is equal to 5/6, so the calculation again confirms that $E(Z) = 5/3$.

Proposition 6.4: For any two independent continuous random variables X and Y, the distribution and density functions for $Z = \max\{X, Y\}$ are given by

$$F_Z(t) = F_X(t) \times F_Y(t) \quad \text{and} \quad f_Z(t) = f_X(t)\, F_Y(t) + f_Y(t)\, F_X(t)$$

The explanation for this observation is simple. To say that $Z \le t$ is to say that $X \le t$ and $Y \le t$, that is, $(Z \le t) = (X \le t) \cap (Y \le t)$, and so

$$F_Z(t) = P(Z \le t) = P(X \le t, Y \le t) = P(X \le t)\, P(Y \le t) = F_X(t)\, F_Y(t)$$

This equation can be differentiated with respect to t to obtain the density function for Z.

6.4 Sums of Random Variables

Often models for common situations lead to consideration of sums of random variables. This means that it is often necessary to know how to determine the probability distribution for a random variable that is described as a sum of two or more other random variables. A few frequently encountered scenarios which involve sums of random variables are listed here.

Scenario 1: An experiment is conducted which results in a numerical reading of some kind. Since experiments, when repeated, often do not produce identical results, it may be desirable to conduct the experiment several times and compare the results of the separate "trials." If the experiment is conducted n times and the n resulting readings are averaged, this numerical average is the random variable

$$X = \frac{1}{n} \times (X_1 + X_2 + \cdots + X_n)$$

where $X_1, \cdots, X_n$ are the random variables that give the readings on the n individual trials.

Often it is desirable to reproduce experiments for the purpose of verification of the results. In such cases it is important to know that the experiments are being conducted "independently" in the sense that the outcome of one experiment doesn't affect the outcome of the others. In this context, the random variables $X_1, \cdots, X_n$ are independent random variables.

Scenario 2: A piece of equipment is put in place, and a "cold standby" backup piece of equipment is available to replace it when it fails. So service is available from the two pieces of equipment for a period as long as the sum of their individual operational lifetimes.

Scenario 3: There are various sources for a particular commodity. (For example, there may be several manufacturers of a particular device.) The total available supply is then the sum of the amounts available from the various sources. If these fluctuate probabilistically, we are then talking about a sum of random variables.

Sums of independent random variables are especially important. One tool for working with such sums is provided by the "convolution" integral.

Definition 6.7: If f and g are functions defined on the entire real number line, then the *convolution* of f and g is the function denoted by $f*g$ and defined by

$$f*g(t) = \int_{-\infty}^{\infty} f(t-s)\, g(s)\, ds$$

Convolution is a commutative operation; that is, $f*g = g*f$. The change of variables $u = t - s$ in the integral demonstrates this fact. (See Problem 6.30.)

The convolution integral is useful in other mathematical settings as well. For example, when Laplace transform techniques are used to solve differential equations, the convolution integral comes into play because the inverse Laplace transform of the product of two functions is the convolution of their inverse Laplace transforms.

In the area of probability, the convolution integral comes into play when sums of independent continuous random variables are encountered.

Proposition 6.5: If X and Y are independent continuous random variables and if $Z = X + Y$, then the density function for Z is given by

$$f_Z = f_X * f_Y$$

Proof: To qualify as a density function for Z, a function f must have the property

$$F_Z(t) = \int_{-\infty}^{t} f(s)\,ds \quad \text{for every real number } t$$

By definition,

$$F_Z(t) = P(Z \leq t) = P(X + Y \leq t) = \iint_{x+y\leq t} f_{X,Y}(x, y)\;dx\,dy$$

$$= \int_{-\infty}^{\infty}\int_{-\infty}^{t-x} f_X(x)\,f_Y(y)\;dy\,dx = \int_{-\infty}^{\infty}\int_{-\infty}^{t} f_X(x)\,f_Y(s-x)\;ds\,dx$$

The last expression is obtained by performing the change of variables $s = x + y$ in the *innermost* of the integrals in the iterated integration. Now, if the order of integration is reversed, we have

$$F_Z(t) = \int_{-\infty}^{t}\int_{-\infty}^{\infty} f_X(x)\,f_Y(s-x)\;dx\,ds$$

The reason that this form is desirable is that now the inner integral is the convolution of f_X and f_Y, which shows that $f_X * f_Y$ satisfies the criteria to be a density function for Z.

Example 6.9. Two random numbers between 0 and 1 are independently produced by a random-number generator. Find the probability distribution of the sum of the two numbers.

Solution: It is intuitively clear that the sum will assume values between 0 and 2. Is it also clear to you that the values of the sum will be more concentrated near 1 than toward the endpoints of the interval?

Denote the two numbers generated independently by X and Y, and denote their sum by $Z = X + Y$.

From Proposition 6.5, we need only compute the convolution of f_X and f_Y, each of which is the uniform density on [0, 1]. This is not difficult if we get the proper geometry in mind. This is shown in the figure below.

Here the functions g and f in parts a and b of the figure are simply two copies of the uniform density on [0, 1]. It is the convolution of g and f that we need to determine. In each part of the figure, the shaded area is the region between the graph and the s axis. In part c we have a similar representation of the function

$f(-s)$, where f is the uniform density on [0, 1]. And in parts d and e we have graphs of $f(t-s)$. In part d the graph of $f(t-s)$ is shown for a typical value of t such that $0 < t < 1$, and part e shows a similar graph for a t value in the interval $1 < t < 2$.

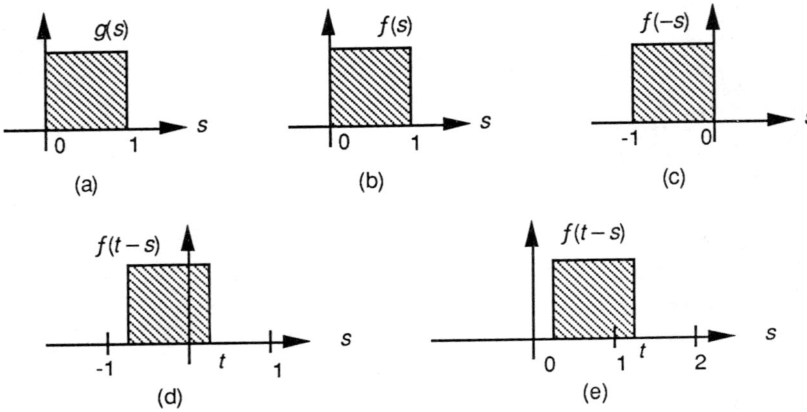

Now the integral to be evaluated is

$$f*g(t) = \int_{-\infty}^{\infty} f(t-s)\, g(s)\, ds$$

The integrand will be nonzero only when *both* factors in the integrand are nonzero. When $0 < t < 1$, this means we should look at the overlap of the shaded regions in parts a and d. This will be the region whose area the convolution integral computes, and we get in that case $f*g(t) = t$ (because the overlap is a rectangle of height 1 and width t). Next, when $1 < t < 2$, we look at the overlap of the shaded regions of parts a and e and see that the area of the overlap now is $2 - t$. [Observe that the base of the overlap rectangle goes from $t - 1$ to 1, so the length of the base of the rectangle is $1 - (t - 1) = 2 - t$.] The same pictures should convince you that $f*g(t) = 0$ if $t \leq 0$ or $t \geq 2$.

Summary: The density function for $Z = X + Y$ is the function

$$f*g(t) = \begin{cases} 0 & t < 0 \\ t & 0 < t < 1 \\ 2 - t & 1 < t < 2 \\ 0 & t > 2 \end{cases}$$

Notice that the result of this calculation does indeed show that values of the random variable $Z = X + Y$ will more concentrated near the center of the interval [0, 2]. (Graph the density function if this isn't obvious to you.)

Proposition 6.6: If X and Y are independent, then

$$E(XY) = E(X)\,E(Y)$$

Proof: We will check the validity of this proposition in the continuous case. The discrete case is similar if one replaces the integrals and the density functions by sums and the mass functions.

Since the product XY is expressed as a simple function of the random variables X and Y, its expected value is given by Proposition 6.3 as

$$E(XY) = \int_{-\infty}^{\infty}\int_{-\infty}^{\infty} xy\, f_{X,Y}(x, y)\; dx\, dy = \int_{-\infty}^{\infty}\int_{-\infty}^{\infty} xy\, f_X(x)\, f_Y(y)\; dx\, dy$$

$$= \int_{-\infty}^{\infty} x\, f_X(x) \int_{-\infty}^{\infty} y\, f_Y(y)\; dy\; dx = \int_{-\infty}^{\infty} x\, f_X(x)\, E(Y)\; dx$$

$$= E(Y) \int_{-\infty}^{\infty} x\, f_X(x)\, dx = E(X)\, E(Y)$$

The key idea in this simple calculation is that the independence of X and Y causes the joint density function to split into a product which is a function of x times a function of y. Once this happens, the double integral splits into a product of two single integrals that evaluate to $E(X)$ and $E(Y)$, respectively.

The converse of Proposition 6.6 is not true. Knowing just the fact that $E(XY) = E(X)E(Y)$ is not enough to conclude that X and Y are independent. There is, in fact, a special term that describes a pair of random variables for which $E(XY) = E(X)E(Y)$. If X and Y satisfy this condition, they are said to be *uncorrelated.* So Proposition 6.6 states that if two random variables are independent, then they are uncorrelated.

Recall that for any pair of random variables X and Y defined on the same sample space, $E(X + Y) = E(X) + E(Y)$. While expected value is additive in this sense, in general variance is not. In the case of independent random variables, however, the variance is additive.

Proposition 6.7: If X and Y are independent, then

$$\text{var}(X+Y) = \text{var}(X) + \text{var}(Y)$$

Proof: If we denote $E(X)$ by μ_X and $E(Y)$ by μ_Y, then

$$E(X+Y) = \mu_X + \mu_Y$$

Therefore from Proposition 3.5 we know that

$$\text{var}(X+Y) = E[(X+Y)^2] - (\mu_X + \mu_Y)^2$$

But then

$$\begin{aligned}\text{var}(X+Y) &= E(X^2 + 2XY + Y^2) - (\mu_X{}^2 + 2\mu_X\mu_Y + \mu_Y{}^2) \\ &= E(X^2) - \mu_X{}^2 + E(Y^2) - \mu_Y{}^2 + 2E(XY) - 2\mu_X\mu_Y \\ &= \text{var}(X) + \text{var}(Y)\end{aligned}$$

since $E(XY) = \mu_X\mu_Y$.

Using mathematical induction, one can extend Proposition 6.7 to any finite collection of independent random variables. If $X_1, \cdots, X_n$ are independent, then

$$\text{var}(X_1 + \dots + X_n) = \text{var}(X_1) + \cdots + \text{var}(X_n)$$

A useful and easy application of this fact is in proving that

$$\sigma_X = \sqrt{npq}$$

whenever X is binomial with parameters n and p. (This was the second part of Proposition 4.1.) Think of X as the number of successes in n trials of an independent trials process, so that

$$X = X_1 + X_2 + \cdots + X_n,$$

where X_k is 0 or 1 depending on whether failure or success occurs on the kth trial. It is easy to see that for $k = 1, 2, \cdots, n$, $\text{var}(X_k) = pq$. The reason is that $E(X_k{}^2) = p$, and that means

$$\text{var}(X_k) = E(X_k{}^2) - E(X_k)^2 = p - p^2 = pq$$

From the additivity of the variance of independent random variables,

$$\text{var}(X) = \text{var}(X_1) + \cdots + \text{var}(X_n) = npq$$

The most "elegant" probability distributions as far as sums of independent random variables are concerned are the binomial distribution and the normal distribution. These are the two that reproduce themselves when independent random variables are summed. Let's consider sums of normal random variables first.

Proposition 6.8: If X and Y are normal random variables and are independent, then $Z = X + Y$ is also normal with

$$\mu_Z = \mu_X + \mu_Y \quad \text{and} \quad \text{var}(Z) = \text{var}(X) + \text{var}(Y)$$

Comment on proof: That $\mu_Z = \mu_X + \mu_Y$ and $\text{var}(Z) = \text{var}(X) + \text{var}(Y)$ is already known; the latter depends on the fact that X and Y are independent. What is new and important here is that Z inherits the distribution shared by X and Y; that is, Z is normal. There are a variety of ways to prove this, all of which are moderately tedious. One way is to use a powerful tool called the *Fourier transform.* For continuous random variables, certain properties can be established by considering the Fourier transforms of the density functions.

Another way to derive the fact that Z is normal is to use the convolution formula of Proposition 6.5. This, however, leads to some quite messy algebra which doesn't shed much light on the proposition under discussion. For that reason we will omit the proof and go on to illustrate its importance in situations to which the proposition applies.

Example 6.10. A component in a system is known to have a normally distributed lifetime with mean 2 years and standard deviation 6 months. A backup component is available to be placed in service when the original fails. The backup is identical to the original. What is the probability that the cumulative service provided by the two components totals more than 5 years?

Solution: Let's consider the length of time the original remains in service to be X and the length of time the replacement remains in service to be Y. The assumption then is that X and Y are normal random variables with mean 2 years and standard deviation 1/2 year.

The combined lifetime of the two then is the random variable $Z = X + Y$, which is normal with mean

$$E(Z) = 2 + 2 = 4$$

years, and variance

$$\text{var}(Z) = \text{var}(X) + \text{var}(Y) = \sigma_X^2 + \sigma_Y^2 = \frac{1}{4} + \frac{1}{4} = \frac{1}{2} \text{ year}$$

Therefore, $\sigma_Z = \sqrt{1/2} \approx .7071$.

Now that we know that Z is normal with $\mu_Z = 4$ and $\sigma_Z = .7071$, we have all the information we need to answer questions about Z. In particular,

$$P(Z \geq 5) = P\left(\frac{Z - \mu_Z}{\sigma_Z} \geq \frac{5 - \mu_Z}{\sigma_Z}\right) = P\left(\frac{Z - \mu_Z}{\sigma_Z} \geq 1.4142\right)$$
$$= 1 - \Phi(1.4142) = .07865$$

(As usual, Φ in this equation denotes the standard normal distribution function. This calculation is based on Proposition 5.3.)

Proposition 6.9: If X and Y are independent binomial random variables with X having parameters n_1 and p and Y having parameters n_2 and p, then $Z = X + Y$ is binomial with parameters $n_1 + n_2$ and p.

Comment on Proposition 6.9: Proposition 6.9 says that the binomial distribution is the discrete distribution that shares the inheritance property that the normal distribution displays. A little reflection on how the binomial distribution arises leads to an intuitively simple understanding as to why it is true. Consider an independent trials process with success probability p in which X counts the number of successes during the first n_1 trials and Y counts the number of successes during the next n_2 trials. The nature of an independent trials process guarantees that X and Y will be independent random variables, and $X + Y$ will give the number of successes during the first $n_1 + n_2$ trials and so will be binomial with parameters $n_1 + n_2$ and p.

Often it is assumed that errors inherent in laboratory procedures cause a normally distributed pattern of results when repeated measurements are made of a given quantity. In fact, the normal density function is sometimes called the "error" function. One technique that is often used in making measurements where errors are expected is to take several measurements and average the results. There is a clear theoretical basis for this. Suppose, for example, that a measurement made in an experiment is assumed to result in a normal random variable with mean μ and variance σ^2. If the experiment is repeated n times independently, we can think of the n resulting measurements as random variables $X_1, \cdots, X_n$, which all have mean μ and variance σ^2 and which are independent. The average Y is then given by $Y = (1/n)(X_1 + \cdots + X_n)$, and the mean and variance of Y are given by

$$\mu_Y = \frac{1}{n}(\mu + \cdots + \mu) = \mu$$

and

$$\text{var}(Y) = \frac{1}{n^2}\sum_{k=1}^{n} \text{var}(X_k) = \frac{1}{n^2} \times n\sigma^2 = \frac{\sigma^2}{n}$$

The fact that the variance of the average of the measurements is appreciably smaller than the variance of an individual measurement means that it is much more probable that the average will be near μ than will an arbitrary single measurement. A common assumption is that the measuring technique is not biased, that is, that the expected value μ of any given measurement agrees with the true value of whatever quantity is being measured.

Example 6.11. The voltage in a circuit is 115 volts. A particular technique for measuring the voltage gives readings which are normally distributed with mean $\mu = 115$ volts and standard deviation $\sigma = 5$ volts. If four readings are taken and the results averaged, the resulting average is normally distributed with mean $\mu = 115$ volts and variance $\sigma^2/4$. So the standard deviation of the average is $5/2 = 2.5$ volts.

Let's do a brief calculation to indicate the value of this. If Y is the average of four readings, then

$$P(|Y - 115| > 3) = P\left(\frac{|Y - 115|}{2.5} > 1.2\right) = 2 - 2\Phi(1.2) = .2301$$

whereas for an individual reading X,

$$P(|X - 115| > 3) = P\left(\frac{|X - 115|}{5} > .6\right) = 2 - 2\Phi(.6) = .5485$$

Summary: The probability that an individual reading will differ from the actual voltage by more than 3 volts is .5485, whereas if four readings are taken and the average used then the probability that the average will differ from from the true value by more than 3 volts is only .2301.

↑ Pre

TEST 2

↓ After (post)

6.5 Conditional Probabilities and Random Variables

In Chapters 1 and 2 we found the concept of conditional probability to be very useful. The idea helps to describe how likely one event is to occur if we know that some other event has occurred.

There is a similar situation regarding pairs of random variables. Suppose we have two discrete random variables X and Y with their associated probability mass functions p_X and p_Y. The mass function p_X carries the necessary information about the probability with which X assumes its various values. But what if we know what value Y assumes? For instance, let's suppose that we know that Y takes on a particular value b. If X and Y are not independent random variables, then in general events of the form $(X = a)$ and $(Y = b)$ are not independent events. This means that $P(X = a)$ and $P(X = a \mid Y = b)$ will not be the same. The conditional probability $P(X = a \mid Y = b)$ is the conditional probability that X takes on the value a given the knowledge that Y takes on value b. This leads to the idea of the *conditional probability mass function*. The conditional probability mass function will be a useful tool for studying the ways in which one random variable influences another.

Definition 6.8: If X and Y are discrete random variables and if b is a number having the property that $P(Y = b) \neq 0$, then the *conditional probability mass function* $p_{X|Y=b}$ is defined by

$$p_{X|Y=b}(x) = P(X = x \mid Y = b)$$

for each real number x.

Recall from the definition of conditional probability that

$$P(X = x, Y = b) = P(Y = b)\, P(X = x \mid Y = b)$$

This equation relates the joint probability mass function of X and Y to the conditional probability mass function $p_{X|Y=b}$, for it can just as well be written as

$$p_{X,Y}(x, b) = p_Y(b)\; p_{X|Y=b}(x)$$

Often situations involving more than one random variable are most naturally described in terms of conditional probability mass functions. The following example illustrates this.

Example 6.12. A particular device is used in various kinds of systems. We will assume that all systems employ either 1, 2, 3, or 4 of these devices and that each of these four possibilities is equally likely to be the case. Each device employed in a system has probability $p = .1$ of failing, and the devices function independently. This means that once we know how many devices are present, the probability distribution of the number of failures will then be known. For example, if it is known that a system employs 3 of the devices, then the number that fail will be binomial with parameters $n = 3$ and $p = .1$.

We will denote by X the number of failures of such devices in the system and by Y the total number of devices employed in the system. What we have observed is that for $b = 1, 2, 3$, and 4, the conditional probability mass function $p_{X|Y=b}$ is the binomial mass function with parameters $n = b$ and $p = .1$.

As a sample computation that can be based on these observations, we will try to determine the probability that there are exactly two failed devices in the system.

Solution:

$$
\begin{aligned}
P(X = 2) &= \sum_{b=1}^{4} P(X = 2, Y = b) \\
&= \sum_{b=1}^{4} P(Y = b)\, P(X = 2 \mid Y = b) \\
&= P(Y = 2)\, P(X = 2 \mid Y = 2) + P(Y = 3)\, P(X = 2 \mid Y = 3) \\
&\quad + P(Y = 4)\, P(X = 2 \mid Y = 4) \\
&= \frac{1}{4}(.1)^2 + \frac{1}{4} C(3,2)\,(.1)^2(.9) + \frac{1}{4} C(4,2)\,(.1)^2(.9)^2 \\
&= .0214
\end{aligned}
$$

Determining the expected value for the number of failed devices in the system is not a difficult computation if we proceed along the lines begun above. Having computed $P(X = 2)$, it is clear that $P(X = 0)$, $P(X = 1)$, $P(X = 3)$ and $P(X = 4)$ can also be determined in a similar manner. Moreover, once all the values of the mass function of X are known, the expected value is easy.

There is another point of view, however, that makes this computation still easier. It involves the idea of *conditional expectation.* Just as the expectation of a discrete random variable utilizes the probability mass function, the conditional expectation utilizes the conditional probability mass function.

Definition 6.9: Suppose that X and Y are discrete random variables and that $P(Y = b) > 0$. The *conditional expectation* of X given that $Y = b$ is denoted by $E(X \mid Y = b)$ and defined by

$$E(X \mid Y = b) = \sum_k x_k P(X = x_k \mid Y = b)$$

where the sum extends over all values x_k assumed by the random variable X.

Notice that all that is happening in this definition is that the probability mass function in Definition 3.6 is now being replaced by the conditional probability mass function.

Proposition 6.10 will now demonstrate a way in which the conditional expectation can be used to determine the unconditional expectation. Depending on what kind of information is available, this may in fact be the most attractive approach to get the expected value. The continuation of Example 6.12 that appears after Proposition 6.10 will illustrate this.

Proposition 6.10: If X and Y are discrete random variables, then

$$E(X) = \sum_j P(Y = y_j)\, E(X \mid Y = y_j)$$

The sum extends over all values y_j assumed by Y.

Proof:

$$\sum_j P(Y = y_j)\, E(X \mid Y = y_j) = \sum_j P(Y = y_j) \sum_k x_k P(X = x_k \mid Y = y_j)$$

$$= \sum_j \sum_k x_k P(Y = y_j)\, P(X = x_k \mid Y = y_j)$$

$$= \sum_j \sum_k x_k P(X = x_k, Y = y_j)$$

$$= \sum_k x_k \sum_j P(X = x_k, Y = y_j)$$

$$= \sum_k x_k P(X = x_k)$$

$$= E(X)$$

Notice that this proof depends only on rearranging the terms of the sum and using elementary properties of conditional probabilities.

Example 6.12 (continued). Proposition 6.10 gives an easy way to determine the expected number of failed devices in the system we were examining in Example 6.12. The conditional probability distribution of X, given the information $Y = 4$, is the binomial distribution with parameters $n = 4$ and $p = .1$. So $E(X \mid Y = 4) = np = 4 \times .1 = .4$. In addition, $E(X \mid Y = 3) = 3 \times .1 = .3$, and $E(X \mid Y = 2) = 2 \times .1 = .2$, and $E(X \mid Y = 1) = 1 \times .1 = .1$. Therefore,

$$E(X) = \frac{1}{4} \times .4 + \frac{1}{4} \times .3 + \frac{1}{4} \times .2 + \frac{1}{4} \times .1 = .25$$

Now let's turn to the case of continuous random variables. What will be meant by the conditional probability density function in the case of two continuous random variables? We cannot proceed as in Definition 6.8 because the two events $(X = x)$ and $(Y = b)$ will both have probability 0 if X and Y are continuous random variables. Remember, however, that even in the discrete case

$$P(X = x \mid Y = b) = \frac{P(X = x, Y = b)}{P(Y = b)}$$

Both the numerator and the denominator of this expression on the right have parallel concepts in the continuous case. The numerator is the joint probability mass function of X and Y, which, of course, corresponds to the joint density function when continuous random variables are considered. Moreover, the denominator is just the mass function for Y, which corresponds to the density function of Y in the continuous case. The analogy then leads to Definition 6.10.

Definition 6.10: Suppose X and Y are continuous random variables. For all numbers y where the density function f_Y satisfies $f_Y(y) \neq 0$, the *conditional density function* of X given that $Y = y$ is denoted by $f_{X|Y=y}$ and defined by

$$f_{X|Y=y}(x) = \frac{f_{X,Y}(x,y)}{f_Y(y)}$$

Similarly,

$$f_{Y|X=x}(y) = \frac{f_{X,Y}(x,y)}{f_X(x)}$$

Example 6.13. A closer look at an example considered previously is in order now. Suppose a point is randomly chosen in the following shaded triangle.

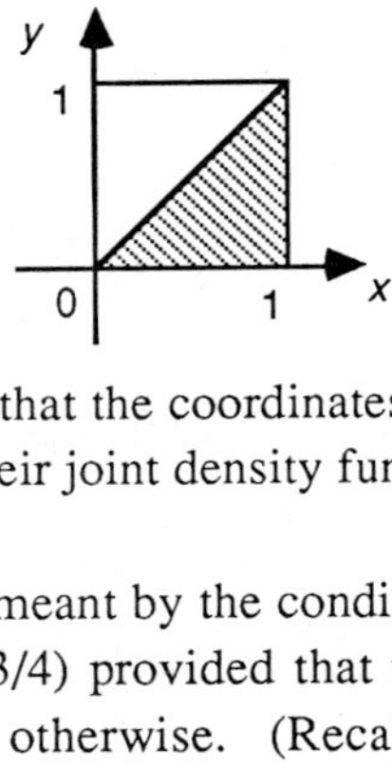

The assumption will be that the coordinates of the point selected, which we denote by X and Y, have as their joint density function the uniform density on the triangle.

In this context, what is meant by the conditional density $f_{X|Y=3/4}$? From the definition, $f_{X|Y=3/4}(x) = 2/f_Y(3/4)$ provided that the point $(x, 3/4)$ lies inside the triangle, and $f_{X|Y=3/4}(x) = 0$ otherwise. (Recall that the joint density has the constant value 2 throughout the triangle and 0 outside the triangle.) Furthermore, the point $(x, 3/4)$ will be inside the triangle precisely when $3/4 < x < 1$.

It is easy to compute (using Proposition 6.2) that $f_Y(y) = 2 - 2y$ whenever $0 < y < 1$. So $f_Y(3/4) = 1/2$. Therefore $f_{X|Y=3/4}(x) = 2 \div .5 = 4$ whenever $3/4 < x < 1$. Notice that this is simply the uniform density function on the interval

[3/4, 1].

One final look at the picture should make all of this fall into place. Once we know that $Y = 3/4$, this guarantees that if (X, Y) is in the triangle then X must satisfy the inequality $3/4 < X < 1$. The "uniformity" of the joint density on the triangle is passed down to X when we are conditioning on the information that $Y = 3/4$ in the sense that the conditional density for X then is the uniform density on the interval [3/4, 1].

Definition 6.11: If X and Y are continuous random variables and $f_Y(y) \neq 0$, then $E(X \mid Y = y)$ denotes the *conditional expectation* of X given that $Y = y$ and is defined by

$$E(X \mid Y = y) = \int_{-\infty}^{\infty} t\, f_{X|Y=y}(t)\; dt$$

Notice that the only way in which the above definition differs from the definition of $E(X)$ is that the density function f_X is replaced by the conditional density function. Also it differs from Definition 6.9 only in that summation of the conditional mass function turns to integration with the conditional density function.

In the continuous case, just as in the discrete case, the conditional expectation can be used to compute the unconditional expectation. The summation and mass function in Proposition 6.10 are replaced by integration and the density function in the continuous case.

Proposition 6.11: If X and Y are continuous random variables having a joint density function, then

$$E(X) = \int_{-\infty}^{\infty} f_Y(y)\, E(X \mid Y = y)\; dy$$

Proof:

$$\int_{-\infty}^{\infty} f_Y(y)\, E(X \mid Y = y)\; dy = \int_{-\infty}^{\infty} f_Y(y) \int_{-\infty}^{\infty} t\, f_{X|Y=y}(t)\; dt\; dy$$

$$= \int_{-\infty}^{\infty} \int_{-\infty}^{\infty} t\, f_Y(y) \frac{f_{X,Y}(t,y)}{f_Y(y)}\, dt\, dy$$

$$= \int_{-\infty}^{\infty} \int_{-\infty}^{\infty} t\, f_{X,Y}(t,y)\, dt\, dy$$

The last integral above is equal to $E(X)$. In fact, this integral is just the special case of the integral in part 2 of Proposition 6.3 in which g is the function $g(x, y) = x$. (You may, of course, change the variable t to x in the double integral here if it helps you to see the connection between this and Proposition 6.3.)

Just as in the discrete case, with models based on continuous random variables, it is often most natural to state assumptions in terms of conditional densities. Example 6.14 illustrates this.

Example 6.14. An electrical cable has a time to failure which is exponentially distributed. The parameter λ in the distribution, however, varies from one manufacturer to another. We will assume that for a given supply of cables of different origins, the parameter in the exponential distribution is uniformly distributed between $\lambda = .5$ and $\lambda = 1$. We will denote by Y the random variable whose value will be the parameter λ in the exponential distribution corresponding to the time to failure for a randomly selected cable from the given supply. And we denote by X the time to failure for such a randomly selected cable, where time is measured in years. This means that for any number y, if we know that $Y = y$, then we know precisely the characteristics of X; that is, the *conditional* density function for X based on the knowledge that $Y = y$ is the exponential density with parameter $\lambda = y$. So in precise mathematical language, the assumptions are that Y is uniform on the interval $[.5, 1]$ and that for a given value of y between .5 and 1, $f_{X|Y=y}$ is the exponential density with parameter $\lambda = y$. This means that the conditional expectation $E(X \mid Y = y)$ will be $1/\lambda = 1/y$.

What is the expected time to failure for a cable randomly selected from this batch?

Solution:

$$E(X) = \int_{-\infty}^{\infty} f_Y(y)\, E(X \mid Y = y)\, dy = \int_{.5}^{1} \frac{2}{y}\, dy = 2 \ln 2 = 1.386$$

Does this make sense? The worst cables (where $\lambda = 1$) have expected time to failure 1 year. The best ones (where $\lambda = .5$) have expected time to failure 2 years. The parameter in the distribution for the available supply of cables is assumed to be uniformly distributed between .5 and 1, so the average would be .75. In other words, $E(Y) = 3/4$. Notice that $E(X \mid Y = 3/4) = 4/3$ since the conditional density $f_{X|Y=3/4}$ is the exponential density with parameter $\lambda = 3/4$. If it seems paradoxical to you that this 4/3 doesn't agree with the answer 1.386 above, the explanation is that two different kinds of averaging are being compared.

Example 6.15. A person will arrive at work between 9 and 10 o'clock in the morning. Sometime before 10 o'clock an important phone call must be placed. Assume that the time of arrival is uniformly distributed between 9:00 and 10:00, and assume that the time that the call is placed is uniformly distributed between the time of arrival and 10:00. What is the probability distribution of the time at which the call is placed, and when is the expected time for the call to be placed?

Solution: This situation ties together several of the concepts of this chapter because there are two random variables that interact. One is the time of arrival (let's call it X), and the other is the time at which the call is placed (which we'll call Y).

If we agree to measure time in hours starting at 9:00, then the assumption regarding X is that X is uniformly distributed on the interval [0, 1]. But what about Y? Here the information is conditional. It is that given the information $X = x$, then Y is uniformly distributed on the interval $[x, 1]$. Or to put it more succinctly, $f_{Y|X=x}$ is the uniform density on the interval $[x, 1]$ for each value of x between 0 and 1.

We can solve for the joint density by simply looking back at Definition 6.10.

$$f_{X,Y}(x, y) = f_X(x)\, f_{Y|X=x}(y)$$

Since $f_X(x) = 1$ whenever $0 < x < 1$ and since $f_{Y|X=x}(y) = 1/(1-x)$ whenever $x < y < 1$, this means that

$$f_{X,Y}(x, y) = \begin{cases} \dfrac{1}{1-x} & \text{if } 0 < x < 1 \text{ and } x < y < 1 \\ 0 & \text{otherwise} \end{cases}$$

The (unconditional) density function for Y can now be calculated from Proposition 6.2. We find that $f_Y(y) = -\ln(1 - y)$ for $0 < y < 1$, with $f_Y(y) = 0$

otherwise. (Check the details of this calculation.)

For the expected value of Y, we can of course compute $E(Y)$ from its own density function:

$$E(Y) = \int_{-\infty}^{\infty} t f_y(t)\, dt = \int_0^1 -t \ln(1-t)\, dt = \int_0^1 (1-t) \ln t\, dt$$

by a simple change of variables. Therefore, using integration by parts we can now derive the expected value.

$$E(Y) = (t - \frac{t^2}{2}) \ln t \,\Big|_0^1 + \int_0^1 \frac{1}{t}(t - t^2/2)\, dt = \frac{3}{4}$$

The first term here is 0. (You have to look carefully at the limit as $t \to 0$ to see that this is true since $\ln t \to -\infty$.) The integral on the right is equal to 3/4. This means that the expected time for the call to be placed is 9:45.

However, if it's the expected value we want, then the easy route is to use Proposition 6.11. Since $f_{Y|X=x}$ is the uniform density on $[x, 1]$, this means that $E(Y \mid X = x) = (x + 1)/2$, simply the midpoint of the interval $[x, 1]$. But then Proposition 6.11 says that

$$E(Y) = \int_{-\infty}^{\infty} f_X(x)\, E(Y \mid X = x)\; dx = \int_0^1 1 \times \frac{x+1}{2}\; dx = \frac{3}{4}$$

6.6 The Central Limit Theorem

Proposition 5.4 is the original version of the central limit theorem and is often referred to as the De Moivre-Laplace theorem. It states that the binomial distribution with parameters n and p is closely approximated by the normal distribution if the parameter n is large.

Recall that every binomial random variable is a sum of Bernoulli random variables. (A Bernoulli random variable is a binomial random variable in which the parameter n is equal to 1.) For example, if Y is the number of successes in 4 trials of an independent trials process, then

$$Y = X_1 + X_2 + X_3 + X_4$$

where X_1 is either 0 or 1, depending upon whether success occurs on the first trial, and X_2, X_3, and X_4 similarly indicate whether success occurs on the second, third, and fourth trials. Furthermore, the random variables $X_1, \cdots, X_4$

all have the same distribution function and they are independent.

If the parameter n in the binomial distribution of Y were large, then Y would be the sum of a large number of independent, identically distributed random variables. A more general form of the central limit theorem says that this is all that is required in order for "convergence" to the normal distribution to take place.

Specifically, suppose that $X_1, X_2, X_3, \cdots$ are random variables which are independent and have a common distribution function. Furthermore, suppose that they have finite mean μ and variance σ^2. (Since they all have the same distribution, they must all have the same mean and variance.) For each positive integer n, let

$$S_n = X_1 + \cdots + X_n \tag{6.2}$$

Then S_n has mean $n\mu$ and variance $n\sigma^2$. (The variance of a sum is the sum of the variances. Proposition 6.7 states this fact for two random variables, but it is equally true for any finite number of independent random variables.)

The De Moivre-Laplace version of the central limit theorem, Proposition 5.4, requires that we subtract the mean of the binomial random variable and divide by the standard deviation in order to make the binomial random variable approximately standard normal. The effect of subtracting the mean and dividing by the standard deviation is to give the "adjusted" random variable mean 0 and variance 1, just as the standard normal distribution has mean 0 and variance 1. The corresponding adjusted form of the random variable S_n in Equation 6.2 is the random variable

$$\frac{S_n - n\mu}{\sigma\sqrt{n}} \tag{6.3}$$

Proposition 6.12 (Central Limit Theorem): If $X_1, X_2, X_3, \cdots$ are independent random variables having the same distribution and having finite mean μ and variance σ^2, and if for each positive integer n

$$S_n = X_1 + \cdots + X_n$$

then

$$\frac{S_n - n\mu}{\sigma\sqrt{n}}$$

is approximately standard normal when n is large. (More precisely, the distribution functions converge to the standard normal distribution function as $n \to \infty$.)

A good model to consider in order to understand the implications of the central limit theorem is that of a numerical-valued experiment conducted over and over again. The result of each experiment can be considered an observation of a random variable. For example, X_1 gives the outcome of the first experiment, X_2 the second, and so forth. If the experiment is conducted in the same manner each time, and if the various experiments are conducted independently of one another, then the random variables $X_1, X_2, X_3, \cdots$ will be independent and will have the same distribution function. However, there is no restriction necessary on just what that distribution might be. For example, each of these random variables could be uniform on [0, 1]. Nevertheless, if the experiment is repeated a large number of times and if S_n is the sum of the results of the first n experiments, then the distribution of the adjusted form of S_n in Expression 6.3 (with mean 0 and variance 1) will be approximately standard normal.

The central limit theorem indicates why the normal distribution is so very important. Independent observations of any type of random variable will "average out" to a normal distribution if the results are averaged in the sense of Equation 6.2 and Expression 6.3.

Problems

6.1 A coin is tossed twice. Let X denote the number of heads on the first toss (0 or 1) and let Y denote the total number of heads on the two tosses (0, 1, or 2). Draw a table similar to Figure 6.1, showing all values of the joint probability mass function for X and Y.

6.2 In Example 6.1, let Z denote the sum of the numbers on the two items selected. Draw a table similar to Figure 6.1, showing all values of the joint mass function for X and Z. As in Figure 6.1, sum the rows and columns to show the values for the mass functions of X and Z along the bottom and right side.

6.3 A device contains three transistors and three resistors. One transistor and one resistor are defective. Two of the six components are randomly selected. Let X denote the number of transistors selected and Y denote the number of defective components selected. Draw a table that shows all the values of the joint probability mass function for X and Y.

6.4 A card is drawn from a standard deck of 52. Let X be the number of hearts drawn and Y the number of red cards drawn. (So X and Y assume values 0 or 1.) Show all values of the joint probability mass function in a table.

6.5 A coin is tossed three times. The random variable X is the number of heads occurring on the first two tosses, and Y is the number of heads occurring on the last two tosses. Draw a table like Figure 6.1 for the joint probability mass function for X and Y. (Observe that X and Y are *not* independent.)

6.6 Let T be the triangle bounded by the x and y axes and the line $x + y = 1$. Suppose that f is the function defined by $f(x, y) = cxy$ for (x, y) in T, and that $f(x, y) = 0$ when (x, y) is not in T.
(a) Find the value that the constant c must have in order for f to be a joint density function.
(b) Suppose that this is the joint density for random variables X and Y. Find the density f_X.
(c) Let $Z = \max\{X, Y\}$. Find $P(Z < 1/2)$.

6.7 Suppose X and Y have the joint density function $f(x, y) = 24xy$ inside the triangle bounded by the line $x + y = 1$ and the coordinate axes, with $f(x, y) = 0$ for points (x, y) not in this triangle. Find the expected values of X, Y, and XY.

6.8 Two random numbers are independently generated using a random number generator which generates numbers according to the uniform density on [0,1]. Find the expected value for the square of the difference of the two numbers.

6.9 Two dice are rolled. Let X denote the maximum of the two numbers that appear, and let Y denote the minimum of the two numbers that appear.
(a) In a table show all values of the joint mass function $p_{X,Y}$.
(b) Find $E(X)$ and $E(Y)$.

6.10 This exercise is designed to check your understanding of what Equation 6.1 says. Suppose two dice are rolled and X and Y are the numbers that appear on the two dice. Let B denote the region in the xy plane consisting of all points (x, y) such that $x^2 + y^2 < 15$. Compute the probability $P\{(X, Y) \in B\}$ and verify that Equation 6.1 is valid in this specific case.

6.11 Suppose X and Y are independent continuous random variables. Show that Definition 6.6 implies that

$$P(a < X < b\,,\, c < Y < d) = P(a < X < b)\, P(c < Y < d)$$

whenever $a < b$ and $c < d$. [Hint: This is elementary; just split the double integral into a product.] Show that this is also true when a or c is $-\infty$ or when b or d is $+\infty$.

6.12 Suppose two random numbers between 0 and 1 are independently generated. (Assume a uniform distribution for each.) What is the probability that the first is more than twice as large as the second? (See Example 6.5.)

6.13 Suppose that X and Y have as joint density function the uniform density on the triangle in the xy plane having vertices at (0, 0), (1, 0), and (1, 1). Determine the density functions of X and Y. Are X and Y independent?

6.14 An experiment is conducted four times. On each attempt, the probability that the experiment succeeds is 1/2. Let X denote the total number of successes, and Y the longest run of successes; that is, the greatest number of successes to occur consecutively during the four experiments. Construct a table showing all values of the joint mass function for X and Y as well as (by summing rows and columns) the mass function of X and the mass function of Y.

6.15 Let $f(x, y) = ye^{-xy-y}$ whenever $x > 0$ and $y > 0$, with $f(x, y) = 0$ otherwise.

(a) Show that f is a density function.

(b) If f is the joint density function for a pair of random variables X and Y, find the density function for X and similarly for Y.

6.16 Complete the calculation of the probability that the device with expected lifetime 8 years lasts more than twice as long as the device with expected lifetime 5 years in Example 6.6.

6.17 A friend says he will call you on the phone between 8 and 9 o'clock. (Assume a uniform distribution for the time of the call during this interval.) Other phone calls occur on your phone according to an exponential distribution; that is, the waiting time for a call is an exponential random variable. Let's assume the expected waiting time for a call from someone other than the friend is 30 minutes. If you arrive at home at precisely 8 o'clock, how likely is it that your friend will be the first person to call you? *Comment on this problem:* Remember the lack-of-memory property that exponential random variables have. This means that if you walk in at 8:00, it really doesn't matter when the last call occurred. You might as well consider the whole process starting from 8:00; that is, the waiting time *as measured from 8:00* until the next call (from someone other than the friend who promised to call) can be assumed to be exponentially distributed with expected value 30 minutes.

6.18 Fill in the details in calculating that $E(Z) = 1/3$ in Example 6.7.

6.19 A room is lighted with two 100-watt bulbs and one 60-watt bulb. During the course of a week, two of the bulbs burn out. Let X be the wattage of the first bulb to burn out and Y the wattage of the second. Assume that bulbs are *not* replaced when they burn out and that the bulbs are equally likely to burn out.

(a) Draw a tree diagram to represent the possibilities.

(b) In a table similar to Figure 6.1, show all values of the joint mass function of X and Y.

(c) What is the expected value of $X + Y$?

6.20 Suppose the random variables X and Y represent the lifetimes of two devices (measured in hours) and that the joint density function for X and Y is given by

$$f(x, y) = .02\, e^{-.1x - .2y}$$

(a) Find f_X and f_Y.

(b) Are X and Y independent?

(c) What is the probability that both devices last longer than 10 hours? [Hint: This is simply $P(X > 10\,,\, Y > 10)$.]

(d) Now find the probability that the sum of the lifetimes of the two devices is greater than 10 hours. [Hint: This is just $P(X + Y > 10)$, but you may find it easier to compute this expression by looking at the complement and using $P(X + Y > 10) = 1 - P(X + Y < 10)$.]

6.21 Answer parts c and d of Problem 6.20 under the assumption that X and Y are independent and uniformly distributed. This time assume that the expected lifetime of each device is 10 hours.

6.22 A device is known to have an exponentially distributed lifetime with expected value 5 years. Your plan is to replace it immediately with a similar device when it stops functioning. So if X and Y are the lifetimes of the devices, then X and Y are independent and each has an exponential distribution with parameter $\lambda = .2$. What is the probability distribution for the length of service you will get from the two devices together? You know, of course, that $E(X+Y) = E(X) + E(Y)$. The question is: What is the probability distribution of $X + Y$? [Hint: Since X and Y are independent, you know their joint density. For any real number t, $F_{X+Y}(t) = P(X+Y \leq t)$. This can be computed by looking at the correct double integral. Once the distribution function is known, differentiate to find the density function for $X + Y$. An alternate solution method would be to compute the convolution of the two exponential densities. See Problem 6.34.]

6.23 Two random numbers between 0 and 1 are independently generated. Assume each is uniformly distributed on [0, 1]. What is the probability distribution and the expected value for the larger of the two numbers obtained? [Hint: Use Proposition 6.4.]

6.24 Two components of a system have exponentially distributed lifetimes, one with expected lifetime 1 year and the other 2 years. Assume the two function independently. What is the probability distribution and the expected value for the length of time until *both* devices have burned out?

6.25 A measurement is made, and the result is a random variable having mean 10 and standard deviation 1. How many times would the experiment have to be repeated independently until the average for all the measurements obtained would have mean 10 and standard deviation less than 1/5?

6.26 Two houses are built in separate flood plains. House A is in a 100-year flood plain, and house B is in a 50-year flood plain. This means that the expected waiting time for a flood in the two areas is, respectively, 100 years and 50 years. Assume these waiting times both to be exponentially distributed random variables. Assume furthermore that these two waiting times are independent random variables. This means that you know the joint density function since you know the distribution of each. What then is the probability that house A will be washed out by a flood before house B is?

6.27 Suppose both X and Y are uniform on [0, 1] and are independent.
(a) Find $P(X + Y < 1)$.
(b) Find $P(X + Y < t)$, assuming that $0 < t < 1$.
(c) Find $P(X + Y < t)$, assuming now that $1 < t < 2$.
(d) Now put all this information together and sketch a graph of the cumulative distribution function for $W = X + Y$.

6.28 Suppose X and Y are random variables having as joint density function the uniform density on the square $S = \{(x, y) \mid 0 < x < 1, 0 < y < 1\}$.
(a) Find the probability $P(Y < X^2)$.
(b) Find $E(XY^2)$.

6.29 In Example 6.8, greater safety (in the sense of a shorter expected time to shutdown) is achieved by putting the two switches in series and having them independently controlled. If they were instead placed in parallel, as in the picture below, the effect would be to make it more difficult to shut down the system.

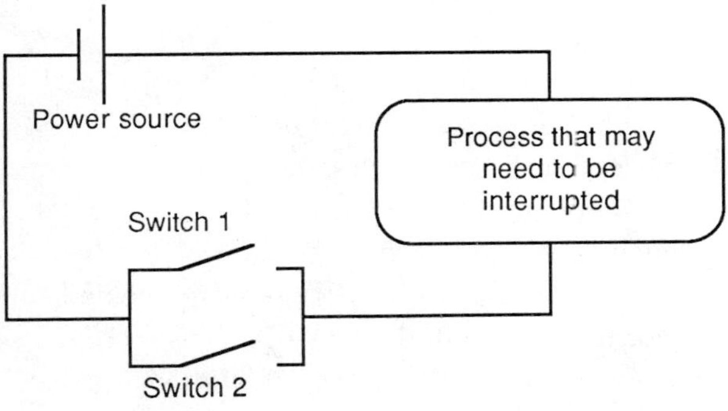

With this kind of configuration, the system would not be shut down until *both* operators had opened their switches. This kind of configuration could be desirable if time is not terribly critical (from the standpoint of the danger involved) and if a shutdown is expensive enough that we elect not to shut down unless both operators agree that a shutdown is necessary. With this configuration, the waiting time for the system to be shut down is the random variable $Z = \max\{X, Y\}$. What is the expected waiting time for the system to be shut down if this configuration is used? (Assume that the assumptions about X and Y remain the same as in Example 6.8.)

6.30 Show that $f*g = g*f$. (See the comment following Definition 6.7.)

6.31 (a) What does the joint probability mass function of X and Y look like if X and Y are the same random variable? Make up a simple example if this situation sounds confusing to you. For example, let X and Y both be equal to the number obtained when a die is rolled. If we were to roll the die twice and let X be the number on the first roll and Y the number on the second, then X and Y are *independent* and certainly not *equal*. (Their probability mass functions *are* equal.) Now, however, you need to think of X and Y as both corresponding to the same roll of the die.

(b) As a follow up to part a, can you intuitively see why in the continuous case if $Y = X$, then it is *impossible* for X and Y to have a joint density function? The reason is that $P\{(X, Y) \in B\}$ would then have to be 0 unless the region B intersects the line $y = x$. From this it can be proved that the joint density would have to be 0 except on the line $y = x$, and such a function could not possibly satisfy

$$\int_{-\infty}^{\infty}\int_{-\infty}^{\infty} f(x,y)\, dx\, dy = 1$$

6.32 Many devices, such as lightbulbs, for example, have time to failure which is approximately normally distributed. (Notice that the exponential distribution is not a good model for a device that is physically wearing out when in use. The lack-of-memory property would be totally out of place here. For a device with the lack-of-memory property, a used device is always just as good as a new one.) Let's suppose that 75-watt lightbulbs have normally distributed time to failure with mean 750 hours and standard deviation 150 hours. If you buy three bulbs, what is the probability that you get more than 2,000 hours use from the three?

6.33 The lifetime of a device is a normal random variable with expected value 1,000 hours and standard deviation 100 hours.

(a) If an identical backup device is available to be placed in service when the original fails, what is the probability that the total length of service they provide is more than 1,800 hours?

(b) If three such devices are available, what is the probability that the cumulative service they provide exceeds 2,700 hours?

(Assume in all cases that the devices function independently; that is, the lifetimes of the devices are independent random variables.)

6.34 Use the convolution formula to derive the distribution of $X + Y$, where X and Y are independent and each has the exponential density with parameter $\lambda = 1$. This computation is not difficult; just be careful about your endpoints of integration.

6.35 A person's blood contains 70 parts per million (ppm) of a certain substance. When a particular technique is used to measure the concentration of the substance in the person's blood, the result is a normal random variable with mean 70 ppm and standard deviation 10 ppm. Assuming that it is possible to reproduce the tests independently, what is the probability that the average of four tests conducted independently would be between 65 and 75 ppm?

6.36 In Example 6.1, find the conditional probability mass functions $p_{X|Y=2}$ and $p_{X|Y=3}$. Then use these to compute $E(X \mid Y = 2)$ and $E(X \mid Y = 3)$.

6.37 If X and Y are independent discrete random variables and if $P(Y = b) \neq 0$, then show that the mass function p_X and the conditional mass function $p_{X|Y=b}$ are identical as functions.

6.38 If X and Y are as in Problem 6.3, find the conditional mass functions $p_{X|Y=1}$ and $p_{Y|X=1}$. Then find $E(X \mid Y = 1)$ and $E(Y \mid X = 1)$.

6.39 In Example 6.14, what is the joint density function for the pair of random variables X and Y?

6.40 In Example 6.4, what is the conditional density function $f_{Y|X=0}$? Now find the conditional expectation $E(Y \mid X = 0)$.

6.41 A random number generator is used to generate 100 random numbers uniformly from the interval [0, 1]. Use the central limit theorem to estimate the probability that the sum of the numbers lies between 50 and 52. [Hint: $S_{100} = X_1 + \cdots + X_{100}$, where the random variables on the right are independent and uniform on [0, 1]. If S_{100} is adjusted so as to have mean 0 and variance 1, the resulting random variable is then approximately standard normal.]

Chapter 7: Stochastic Processes

Sometimes a situation is best described by a large family of related random variables rather than by a single random variable or a few jointly distributed ones. A *stochastic process* is a family of random variables indexed on some set of real numbers. One common situation is that of a sequence $X_1, X_2, \cdots$ of random variables. In this case the subscripts are positive integers. Another common situation is to have a family of the form $\{X_t\colon t \geq 0\}$, in which case the subscripts may be any non-negative real numbers. The former is an example of a *discrete parameter stochastic process* since the positive-integer subscripts form a discrete set of numbers. The latter is a *continuous parameter stochastic process* because the parameter (subscript) t is measured on a continuous scale. In the most common examples of continuous stochastic processes, t represents time.

7.1 Independent Trials Processes

Without using the language of stochastic processes, we have already repeatedly encountered one of the most common ones. An *independent trials process* is an example of a stochastic process. Here the random variables of interest are the random variables $X_1, X_2, \cdots$ where, for each positive integer n, X_n is the number of successes to occur during the first n trials of the independent trials process. The entire process cannot be described via a single random variable if the number of trials is not fixed in advance, but rather the interaction of the random variables describes the process.

In an independent trials process, it is easy to see that the only way that there can be exactly k successes in the first n trials is for one of two things to happen. Either there must be k successes during the first $n - 1$ trials and failure on the nth trial, or else there must be $k - 1$ successes during the first $n - 1$ trials and success on the nth trial. This leads to the observation stated next as Proposition 7.1. Equation 2 of the proposition is just an application of the familiar multiplicative law from Proposition 1.1 in Chapter 1.

Proposition 7.1: In an independent trials process with success probability p, if X_n denotes the number of successes to occur during the first n trials, then the random variables $X_1, X_2, \cdots$ are related as follows:

1. $P(X_1 = 1) = p, P(X_1 = 0) = 1 - p$ and

2. $P(X_n = k) = p\,P(X_{n-1} = k - 1) + (1 - p)\,P(X_{n-1} = k)$

It is, in fact, true that Equations 1 and 2 in Proposition 7.1 characterize the binomial distribution. In other words, if all one knows about $X_1, X_2, \cdots$ is that Equations 1 and 2 in Proposition 7.1 are valid, it is easy to show that in fact each of the random variables X_n does have the binomial distribution with parameters n and p. This argument uses mathematical induction. Suppose that Equations 1 and 2 are known to be true. Equation 1 says that X_1 is binomial with parameters $n = 1$ and p. Assume now that X_{n-1} is binomial with parameters $n - 1$ and p. Then from Equation 2,

$$P(X_n = k) = p\,P(X_{n-1} = k - 1) + (1 - p)\,P(X_{n-1} = k)$$

But since X_{n-1} is binomial with parameters $n - 1$ and p, this means that

$$\begin{aligned} P(X_n = k) &= p\,C(n - 1, k - 1)\,p^{k-1}\,(1 - p)^{n-k} + (1 - p)\,C(n - 1, k)\,p^k\,(1 - p)^{n-1-k} \\ &= C(n - 1, k - 1)\,p^k\,(1 - p)^{n-k} + C(n - 1, k)\,p^k\,(1 - p)^{n-k} \\ &= \{C(n - 1, k - 1) + C(n - 1, k)\}\,p^k\,(1 - p)^{n-k} \\ &= C(n, k)\,p^k\,(1 - p)^{n-k} \end{aligned}$$

The last line uses the fact that $C(n - 1, k - 1) + C(n - 1, k) = C(n, k)$. There is an easy way to see why this little identity is true by just remembering that $C(n, k)$ is the number of ways of choosing k elements from a set of n elements. If we separate out one object from a set of n objects, then we can choose k objects from the entire group by either (1) choosing k objects from the $n - 1$ still grouped together, or (2) choosing $k - 1$ from the $n - 1$ and then including the one sitting off to the side. Since (1) can be performed in $C(n - 1, k)$ different ways and (2) can be performed in $C(n - 1, k - 1)$ different ways, this gives $C(n - 1, k) + C(n - 1, k - 1)$ ways of choosing k objects from the n objects.

7.2 A One-Dimensional Random Walk

Example 7.1. A computer network serves many users simultaneously. Each time there is a change in the number of current users, we will suppose that the probability is p that a new user is entering the system and that the probability is $q = 1 - p$ that an old user is leaving the system.

Question: What is the probability distribution for the net gain in the number of users of the system after a given number of users have either entered or left the system?

Model: If we think of Y_n as the net gain after n "additions" or "subtractions" from the system, then the random variables $Y_1, Y_2, \cdots$ form the stochastic process of interest. Though they do not have a binomial distribution, they are closely related to a sequence of binomial random variables. Denote by X_n the number of "additions" to the system during the first n changes. This means that $n - X_n$ will be the number of "subtractions" from the system during these n changes. The advantage in looking at X_n rather than Y_n is that X_n *is* binomial with parameters n and p, and Y_n can now be described in terms of X_n. For in fact the net gain Y_n after n changes in the number of users on the system will be equal to the number of new additions minus the number of people leaving the system. The net increment then is $X_n - (n - X_n) = 2X_n - n$. Thus for a given integer k,

$$P(Y_n = k) = P(2X_n - n = k) = P\left(X_n = \frac{n+k}{2}\right)$$

One thing to notice is that the right side here is clearly 0 unless $n + k$ is an even integer. This is because after an even number of changes in the number of users (that is, when n is even), the net gain must be an even number (that is, Y_n is even). Thus $P(Y_n = k)$ will necessarily be 0 if n is even and k is odd. For similar reasons, $P(Y_n = k)$ will be 0 if n is odd and k is even. Since X_n is known to be binomial with parameters n and p, we can now describe the probability distribution of Y_n.

If $\alpha = (n + k)/2$ is an integer, then

$$P(Y_n = k) = P(X_n = \alpha) = C(n, \alpha)\, p^{\alpha} q^{n-\alpha}$$

For instance, suppose that the number of users of the system initially was 10, that the number of users has changed (because of persons joining or leaving the system) 5 times, and that each time there is a change the probability is 2/3 that someone is entering the system and 1/3 that someone is leaving. Then,

$$P(Y_5 = 3) = P\left(X_5 = \frac{5+3}{2}\right) = P(X_5 = 4)$$

$$= C(5, 4)\, p^4 q = 5\, p^4 q = 5\left(\frac{2}{3}\right)^4\left(\frac{1}{3}\right) = \frac{80}{243}$$

The probability that there will be exactly 13 users after there have been exactly 5 changes in the number of users on the system is 80/243.

We should pause to acknowledge that, as is often the case with mathematical models, this model requires that certain restrictions be placed on the inputs if we want to represent a realistic situation. The restriction is that a negative number of users of the computer network is meaningless. This means that the computation performed actually gives the correct probabilities for the net gain or loss to the system only as long as the number of users is always positive. For example, if the system had 4 users to start with, then the fact that $P(Y_5 = 3) = 80/243$ does *not* mean that after 5 changes in the number of users the probability is 80/243 that there will be 8 users. The reason is that if the first four "occurrences" are people leaving the system, at that point the number of users has dropped to 0. At that point the assumptions we used in the model become invalid because it is impossible for anyone else to leave the system. Many mathematical models, whether they are deterministic (as in differential equations models) or probabilistic, are only approximations of real phenomena and are valid for only a restricted range of inputs. The model of this example is accurate as long as the number of users is larger than the number of people entering and leaving.

7.3 Poisson Processes

While the Poisson and exponential distributions have been introduced earlier, it is in the context of modeling phenomena via the concept of a Poisson process that the intricate relationship between Poisson and exponential random variables is most visible.

At an intuitive level, the essence of a Poisson process is the idea of "random phenomena" occurring intermittently in accordance with certain descriptive assumptions. Real phenomena that might be modeled via a Poisson process are things such as the following: (1) calls arriving at a telephone switchboard, (2) breakdowns of a piece of equipment, (3) traffic entering a parking lot, (4) emissions of alpha particles from a quantity of radioactive substance, and (5) traffic accidents in a city. Some of these examples are idealized. For example, if a traffic light affects the flow of traffic into a parking lot, the movement of cars into the lot will not satisfy the axioms of a Poisson process.

A Poisson process involves a family $\{X_t : t > 0\}$ of random variables. The basic idea is that for a given time t, X_t counts the number of occurrences during the time interval $[0, t]$ of whatever phenomena is being modeled. So time is measured on a continuous scale relative to some fixed "starting time" referred to as time $t = 0$. For example, if we are observing cars entering a parking lot and if time is being measured in minutes, then X_{10} would be the number of cars that enter the parking lot during the first 10 minutes. Notice that $X_{t_2} - X_{t_1}$ would represent the number of occurrences of the phenomena *between* times t_1 and t_2.

Let's look at the precise requirements that a Poisson process must satisfy and try to understand intuitively what each of the conditions is all about. The requirements are given in Definition 7.1.

Definition 7.1: A *Poisson process* is a family $\{X_t : t \geq 0\}$ of non-negative integer-valued random variables related in the following ways:

1. $X_0 = 0$, and, if $t_1 \leq t_2 \leq t_3 \leq t_4$, then $X_{t_2} - X_{t_1}$ and $X_{t_4} - X_{t_3}$ are independent random variables.
2. There is a constant $\mu > 0$ such that for any $t > 0$,

(a) $$\lim_{\Delta t \to 0} \frac{1 - P(X_{t+\Delta t} = X_t)}{\Delta t} = \mu$$

(b) $$\lim_{\Delta t \to 0} \frac{P(X_{t+\Delta t} - X_t > 1)}{\Delta t} = 0$$

(c) $$\lim_{\Delta t \to 0} \frac{P(X_{t+\Delta t} - X_t = 1)}{\Delta t} = \mu$$

Property 1 simply says that we start counting at time $t = 0$ and that the number of occurrences in two disjoint time intervals should be independent of each other. This is clearly intuitively plausible for many situations. If we are modeling traffic accidents, the number of accidents between 10 o'clock and 11 o'clock would have no apparent reason to affect the number of accidents between 1 o'clock and 2 o'clock.

Property 2(c) says that for small time intervals, the probability of exactly one occurrence of the phenomena during the time interval is approximately proportional to the length of the time interval. The constant μ is the constant of proportionality.

Property 2(b) says that for short time intervals the likelihood of more than one occurrence is negligible, that is, negligible in comparison to the length of the time interval.

Notice that the expression $1 - P(X_{t+\Delta t} = X_t)$ in the numerator of Property 2(a) is 1 minus the probability that no occurrence has taken place during the time interval from t to $t + \Delta t$. But since $1 - P(E) = P(E^c)$ for any event E, this is simply the probability of at least one occurrence between time t and time $t + \Delta t$. So Property 2(a) says that for small time intervals the probability of at least one occurrence is approximately proportional to the length of the time interval, with μ being the constant of proportionality. Property 2(a) is a logical consequence of Properties 2(b) and 2(c), but it is listed separately for reference later.

Investigation of the Poisson Process

If $\{X_t: t \geq 0\}$ is a Poisson process as described by the above properties, it is interesting to investigate the properties of the function $f_0(t)$ defined for all $t \geq 0$ by

$$f_0(t) = P(X_t = 0)$$

that is, $f_0(t)$ is the probability that no occurrences have yet taken place at time t. Then,

$$f_0(t + \Delta t) = P(X_{t+\Delta t} = 0) = P(X_t = 0, X_{t+\Delta t} = 0) = P(X_t = 0)\, P(X_{t+\Delta t} = 0)$$

Being able to write this last product depends on Property 1; that is, the random variables X_t and $X_{t+\Delta t} - X_t$ are independent. But now

$$\frac{f_0(t + \Delta t) - f_0(t)}{\Delta t} = \frac{P(X_t = 0)\, P(X_{t+\Delta t} = X_t) - P(X_t = 0)}{\Delta t}$$

$$= P(X_t = 0)\frac{P(X_{t+\Delta t} = X_t) - 1}{\Delta t} \to -\mu f_0(t) \text{ as } \Delta t \to 0$$

[This uses Property 2(a).] However, the value of this limit is, by definition, the derivative $f_0'(t)$, and so we have shown that $f_0'(t) = -\mu f_0(t)$.

This simple differential equation is easily solved to give $f_0(t) = Ae^{-\mu t}$, where A is constant. If we take into account now that $f_0(0) = P(X_0 = 0) = 1$ (because no occurrences have occurred yet at time $t = 0$), this tells us that $A = 1$, and so

$$f_0(t) = P(X_t = 0) = e^{-\mu t} \tag{7.1}$$

Now we need to consider the probabilities of X_t assuming positive values. For each positive integer k, let's denote by f_k the function

$$f_k(t) = P(X_t = k)$$

that is, $f_k(t)$ is the probability that exactly k occurrences have taken place by time t. We will need to make use of the following decomposition for $f_k(t+\Delta t)$:

$$\begin{aligned} f_k(t+\Delta t) &= P(X_{t+\Delta t} = k) \\ &= P(X_t = k, X_{t+\Delta t} - X_t = 0) + P(X_t = k-1, X_{t+\Delta t} - X_t = 1) \\ &\quad + P(X_t = k-2, X_{t+\Delta t} - X_t = 2) + \ldots + P(X_t = 0, X_{t+\Delta t} - X_t = k) \\ &= P(X_t = k)\, P(X_{t+\Delta t} - X_t = 0) + P(X_t = k-1)\, P(X_{t+\Delta t} - X_t = 1) \\ &\quad + P(X_t = k-2)\, P(X_{t+\Delta t} - X_t = 2) + \cdots + P(X_t = 0)\, P(X_{t+\Delta t} - X_t = k) \end{aligned}$$

Therefore $\dfrac{f_k(t+\Delta t) - f_k(t)}{\Delta t}$ is given by

$$\frac{1}{\Delta t} \times \{\, P(X_t = k)\, P(X_{t+\Delta t} - X_t = 0) + P(X_t = k-1)\, P(X_{t+\Delta t} - X_t = 1)$$

$$+ \cdots + P(X_t = 0)\, P(X_{t+\Delta t} - X_t = k) - P(X_t = k) \,\}$$

$$= P(X_t = k)\, \frac{P(X_{t+\Delta t} - X_t = 0) - 1}{\Delta t} + \frac{P(X_t = k-1)\, P(X_{t+\Delta t} - X_t = 1)}{\Delta t}$$

$$+ \text{ other terms}$$

where the sum of the other terms is dominated in absolute value by

$$\frac{P(X_{t+\Delta t} - X_t > 1)}{\Delta t}$$

which converges to 0 as $\Delta t \to 0$, by Property 2(b).

Taking the limit as $\Delta t \to 0$ gives us the information that

$$f_k'(t) = -\mu f_k(t) + \mu f_{k-1}(t) \tag{7.2}$$

Since we already know what f_0 is, we can use Equation 7.2 with $k = 1$ to find f_1. The function f_1 must satisfy the differential equation

$$f_1'(t) = -\mu f_1(t) + \mu\, e^{-\mu t}$$

This differential equation is easily solved to give $f_1(t) = (\mu t + c)e^{-\mu t}$, where c is a constant. However, $f_1(0) = P(X_0 = 1) = 0$, since there have been no occurrences yet at time $t = 0$, and this tells us that the constant $c = 0$, which allows us to conclude that

$$f_1(t) = \mu t e^{-\mu t}$$

Now, using Equation 7.2 again, it is possible to compute f_2, since we now know f_1. Equation 7.2 says that

$$f_2'(t) = -\mu f_2(t) + \mu f_1(t) = -\mu f_2(t) + \mu\{\mu t e^{-\mu t}\}$$

This differential equation, together with the initial condition $f_2(0) = P(X_2 = 0) = 0$, can be easily solved as a first-order linear differential equation to obtain

$$f_2(t) = \frac{1}{2}\,\mu^2\, t^2\, e^{-\mu t}$$

The method that has been used to find f_1 and f_2 from Equation 7.2 can be continued, since each function f_k is recursively defined in terms of f_{k-1} by Equation 7.2. It is not difficult to show via mathematical induction that for any positive integer k,

$$f_k(t) \;=\; P(X_t = k) \;=\; \frac{1}{k!}\,(\mu t)^k\, e^{-\mu t} \tag{7.3}$$

Recall now from Chapter 4 the definition of the Poisson distribution. A random variable X is a Poisson random variable with parameter λ provided

$$P(X = k) \;=\; \frac{1}{k!}\, e^{-\lambda}\, \lambda^k$$

The computation we have just finished has demonstrated Proposition 7.2.

Proposition 7.2: If $\{X_t : t \geq 0\}$ is a Poisson process as in Definition 7.1, then for each positive number t, X_t is a Poisson random variable with parameter

$$\lambda = \mu t$$

The constant μ in Definition 7.1 is referred to as the *average intensity* or simply the *intensity* of the process. The reason is that μ represents the average number of occurrences of the phenomena per unit of time. To see this simply remember that for a Poisson random variable X, $E(X) = \lambda$, where λ is the parameter in the distribution. (See Problem 4.9.) Therefore, since X_t has a Poisson distribution with parameter $\lambda = \mu t$, it follows that $E(X_t) = \mu t$; that is, the expected number of occurrences of the phenomena being studied during the time interval $[0, t]$ is μt, the intensity of the process times the length of the time interval.

Example 7.2. Suppose that the calls coming into a switchboard constitute a Poisson process with intensity $\mu = 4$ calls per minute. Then for any value of t, the random variable X_t, which counts the number of calls that have arrived by time t, has the Poisson distribution with parameter $\lambda = 4t$. So if we wish to measure the number of calls coming into the switchboard during a 10-minute time period, we can think of this as a Poisson process in which X_{10} is the random variable of interest, and it will be Poisson with the parameter $\lambda = 4 \times 10 = 40$. Since the probability distribution is now explicitly known, any probabilities of interest can be computed. For example, the probability that 30 or fewer calls come in during a 10-minute time interval would be

$$P(X_{10} \leq 30) = \sum_{k=0}^{30} e^{-40} \frac{40^k}{k!}$$

Connection between the Poisson Process and Exponential Distribution

As we have just seen, if we count the number of occurrences of the phenomena being observed in a Poisson process during a fixed time interval, the random variable that does the counting has a Poisson distribution. Instead of counting the number of occurrences during a fixed time interval, what if we consider the waiting time for the first occurrence in a Poisson process? Let Y_1 be this random variable; that is, Y_1 gives the length of time from the starting time (time $t = 0$) to the time of the first occurrence. In Example 7.2, Y_1 would be the length of time that passes during the observed time period before the first call comes in to the switchboard.

The key relation between Y_1 and the Poisson process $\{X_t : t \geq 0\}$ is that $P(Y_1 > t) = P(X_t = 0)$. This is because $(Y_1 > t)$ and $(X_t = 0)$ represent precisely the same event; to say that the waiting time for the first occurrence is greater than t is the same as saying that at time t there have been no occurrences. However, Equation 7.1 gives $P(X_t = 0)$ as $e^{-\mu t}$. Therefore $P(Y_1 > t) = e^{-\mu t}$ and $P(Y_1 \leq t) = 1 - e^{-\mu t}$. We can differentiate to find that for $t > 0$, the density function of Y_1 is the exponential density $f_{Y_1}(t) = \mu e^{-\mu t}$ for $t > 0$. *So Y_1, the waiting time for the first occurrence, is exponential with parameter μ.*

It might be a good idea at this time to review the analogy between an independent trials process and a Poisson process. The binomial distribution arises when we fix the number of trials in an independent trials process and count the number of successes. The Poisson distribution arises when we fix the time interval in a Poisson process and count the number of occurrences of the observed phenomena. The geometric distribution arises in an independent trials process when we consider "waiting time for first success" in terms of the number of trials required. The exponential distribution arises when we consider "waiting time for first occurrence" in a Poisson process in which time is measured on a continuous scale. Both of these "waiting time" phenomena, described by geometric or exponential random variables, have the lack-of-memory property discussed in Chapters 4 and 5.

Proposition 7.3 summarizes the main features of a Poisson process. Property 3 says, for example, that if we let Y_2 denote the length of time required for *two* occurrences (in other words the total waiting time from time $t = 0$ until the *second* occurrence), then $Y_2 - Y_1$ has the same distribution as Y_1. What is of perhaps more interest is that $Y_2 - Y_1$ is independent from Y_1. All this is summarized by saying that waiting time between the first and second occurrence is independent of the time of the first occurrence and has the same probability distribution as the waiting time for the first occurrence. This is a consequence of the lack-of-memory property. In fact it can be shown that the entire Poisson process has a certain lack-of-memory feature in that the choice of starting time is irrelevant. In other words, for any $t_0 > 0$, the random variables X_t and $X_{t_0+t} - X_{t_0}$ have the same distribution. The former counts occurrences during the time interval $[0, t]$ and the latter during the interval $[t_0, t_0 + t]$. The distributions turn out to be the same because the intervals are of the same length. A stochastic process having this property is said to have *stationary increments*. There are other senses in which some stochastic processes are unaffected by time shifts. *Stationary stochastic processes* are introduced and discussed in Section 7.7.

Proposition 7.3 (Properties of Poisson Processes): Suppose some kind of random phenomena occur intermittently. For each positive integer t, let X_t denote the number of occurrences during the time interval $[0, t]$. Also, for each positive integer n, let Y_n denote the time at which the nth occurrence takes place.

If $\{X_t\}$ is a Poisson process with parameter μ, then all the following are true:

1. For every $t_0 > 0$, the random variable $X_{t_0+t} - X_{t_0}$ (which gives the number of occurrences during the time interval from t_0 to $t_0 + t$) is a Poisson random variable with parameter $\lambda = \mu t$.

2. The number of occurrences during disjoint time intervals constitute independent Poisson distributed random variables.

3. The random variables, $Y_1, Y_2 - Y_1, Y_3 - Y_2, \cdots$ are all exponentially distributed with parameter μ. This says that the waiting time for the first occurrence, the waiting time between the first and the second occurrence, between the second and the third, and so on, are all exponential with parameter μ. Furthermore, these random variables are mutually independent.

4. The constant μ can be interpreted as the *average number of occurrences per unit time*.

5. If there is exactly one occurrence between times t_1 and t_2, then the time of that occurrence is uniformly distributed on the interval $[t_1, t_2]$.

7.4 Poisson or Binomial?

Example 7.3. Suppose a snowstorm has begun and a grid such as the one shown in Figure 7.1 has been laid out on the ground. The "intensity" of the snowfall is 3 flakes per square inch per minute. The experiment we are going to do is to count the number of snowflakes hitting the one-square-inch shaded section of the grid during a specified 1-minute time interval. Let's call this random variable X. What kind of model is appropriate here?

Solution: The first thing to consider is that there are at least two different interpretations to the statement that "snowflakes are falling at a rate of three snowflakes per square inch per minute."

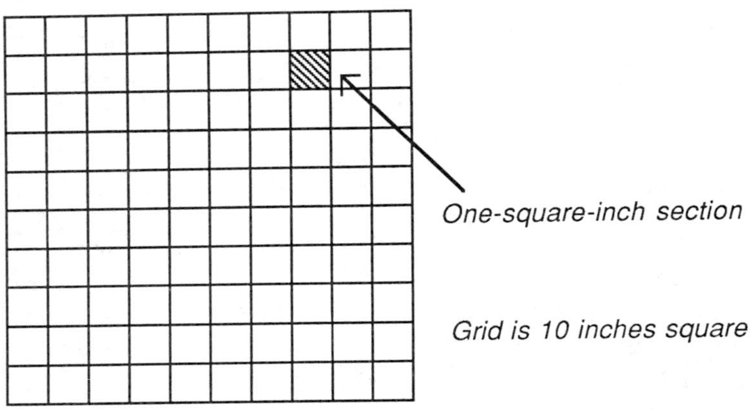

Figure 7.1 Snowflakes are falling on this grid at a rate of three snowflakes per square inch per minute.

Model 1: Suppose we interpret this as meaning that 300 snowflakes fall on the entire 100-square-inch grid during each minute. We can think then of each snowflake as an independent trial that may or may not fall on the one square inch section where we are doing the count. All sections being the same size, we have $p = .01$ as the probability that a given snowflake hits the shaded square. From this point of view, the number of snowflakes hitting the shaded square during a 1-minute time interval is binomial with $n = 300$ and $p = .01$. The expected value is $np = 3$, which is, of course, to be expected (no pun intended) since the average is

3 snowflakes per square inch per minute.

Let's do a simple computation using this model. What is the probability that exactly two snowflakes hit the shaded square during the observed minute?

$$P(X = 2) = C(300, 2)\, p^2 q^{298} = C(300, 2)\, .01^2\, .99^{298} \approx .2244$$

Model 2: Now let's model this situation as a Poisson process. Since the intensity of the snowfall is three snowflakes per square inch per minute, this means that X, the number of snowflakes to hit the shaded square in a 1-minute interval, will now be considered to have a Poisson distribution with parameter λ = intensity $\times$ length of time interval $= 3 \times 1 = 3$. Since X is Poisson now, $E(X) = \lambda$, by Problem 4.9, so again we have $E(X) = 3$, just as it should be.

In this model, what is the probability that exactly two snowflakes hit the shaded square during the 1-minute interval?

$$P(X = 2) = e^{-\lambda} \frac{\lambda^2}{2!} = e^{-3} \times \frac{9}{2} \approx .2240$$

Isn't it astounding that two apparently very different models should give such nearly identical answers for this calculation? The reason this happens can be explained mathematically by saying that if λ is fixed, then the binomial distribution with parameters n and $p = \lambda/n$ converges to the Poisson distribution with parameter λ as $n \to \infty$. A less precise but intuitively more pleasant way of describing this is to say that if n is "large" and p is "small," then the binomial distribution with parameters n and p is approximately the same as the Poisson distribution with parameter $\lambda = np$.

But which model *is* the correct one? A little further insight is gained by considering one more scenario. Suppose that the 10 by 10 grid is replaced by a 100 by 100 grid and that the binomial model is used. Now $n = 30{,}000$ and $p = .0001$. However, $np = 3$ still. In this model,

$$P(X = 2) = C(30000, 2)\, .0001^2\, .9998^{29998} \approx .2240$$

This calculation is consistent with the Poisson model to at least four significant digits. Furthermore, it sheds additional light on the question as to which model is correct. If one interprets the average rate of 3 snowflakes per square inch per minute as being an exact rate measured over a finite grid, then the correct model is a binomial model based on the idea of an independent trials process. If, however, one imagines an infinite plane on which snow is falling at a rate of 3 snowflakes per square inch per minute, then the correct model is the Poisson model based on the idea of a Poisson process. (Of course, the very idea of snowflakes falling at a rate of 3 per square inch per minute over an infinite plane inherently involves the limit concept.)

7.5 Sample Functions

The mathematical definition of a stochastic process is precise enough. It is useful also, however, to know a few productive ways to think about them. This is especially true when the parameter t in the process $\{X_t\}$ varies over a continuum of real numbers rather than a discrete set. Let's illustrate with an example.

Example 7.4. Consider a stochastic process of the form

$$X_t = A \cos(\omega t + \theta)$$

where ω is constant, but where A and θ are random variables. Suppose for definiteness that A is uniformly distributed on the interval $[0, 1]$ and θ is uniformly distributed on the interval $[0, 2\pi]$. Since ω is a constant, the frequency of the wave is absolutely determined. The amplitude and phase angle are probabilistic, however, and will depend on the values of the random variables A and θ. Each pair of values for A and θ ($0 \le A \le 1$ and $0 \le \theta \le 2\pi$) leads to a unique curve. For example, the special case of $A = .3$ and $\theta = 2.6$ would give one possible *observation* of this process. Common terminology is to call each of these possible functions a *sample function*. In a sense you can think of the sample functions as the *possible outcomes* of the process being observed. This is an extension of the way in which we used the term *possible outcomes* in Chapter 1 to refer to the possible results of a simple experiment with a finite sample space.

It is also worthwhile to compare the way one looks at a stochastic process $\{X_t\}$ to the way one looks at values of a single random variable. Suppose Y is a single random variable defined on a sample space S. Then for every possible outcome $s \in S$, $Y(s)$ is simply a real number. Now in the case of a stochastic process $\{X_t\}$, each of the random variables X_t is defined on a common sample space S, so for each $s \in S$ and each t, $X_t(s)$ is a real number, the observed value of X_t. Often it is useful to think of s as fixed and to consider how $X_t(s)$ looks as a function of t. For each fixed $s \in S$, the function $t \to X_t(s)$ is simply a real-valued function of a real variable.

To make this idea still more concrete, let's return to the example

$$X_t = A \cos(\omega t + \theta)$$

Here ω is a fixed constant. We'll pretend that the values of the random variables A and θ are produced independently by random-number generators so that $0 \le A \le 1$ and $0 \le \theta \le 2\pi$. The "possible outcomes" of the experiment then correspond to pairs of random numbers produced, that is, the values of A and θ. For each such "possible outcome," there results a sample function. The sample function

$$f(t) = .2 \cos(\omega t + 5)$$

being the result, for instance, if the values produced by the random-number generator for A were .2 and for θ were 5.

The stochastic process $X_t = A \cos(\omega t + \theta)$ is somewhat specialized in the sense that every random variable X_t here is described in terms of the random variables A and θ. Generally the random variables that make up a stochastic process do not have such a simple relationship connecting them with each other. Even so, the concept of a "sample function" remains that of fixing an outcome $s \in S$ in the underlying sample space and considering how $X_t(s)$ varies as a function of t.

As a further illustration, suppose we are monitoring a Poisson process such as calls coming into a telephone switchboard. One "possible outcome" can be partially described by saying that the first call arrives at time $t_1 = 1.3$ minutes, the second at time $t_2 = 3.2$ minutes, and the third at time $t_3 = 3.8$ minutes. In this case, a portion of the graph of the sample function that corresponds to such an "observation" of the process would be as shown in Figure 7.2. This is a graph of X_t corresponding to the observation just partially described, where X_t is the total number of calls to have arrived by time t.

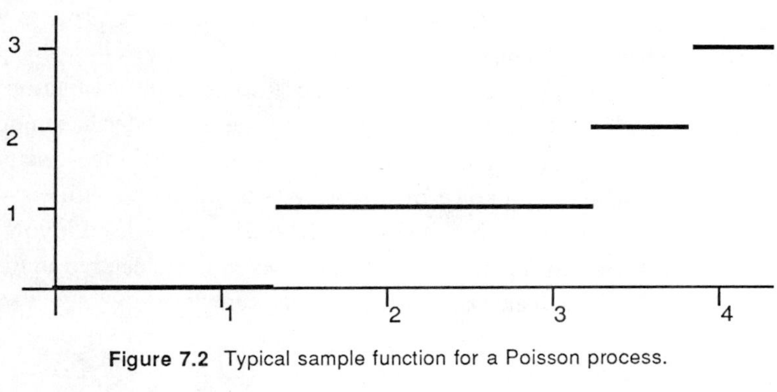

Figure 7.2 Typical sample function for a Poisson process.

There are some very subtle points related to the study of stochastic processes that require sophisticated and advanced mathematics. This can be illustrated, for example, even with regard to an independent trials process. How does one think of the sample space for an infinite sequence of coin tosses? A finite sequence of tosses is no problem, but when one envisions the stochastic process $\{X_n\}$, where X_n is the number of heads to occur in the first n tosses of an infinite sequence of tosses, things get a bit confusing. While it's clear what the probability distribution of X_n is going to be, it's not so clear just how one can envision a sample space (with probability measure) on which all of the X_n's are defined. Fortunately, one can do a lot of worthwhile modeling without dealing with such riddles, but it is worth realizing that the complexities required to do a rigorous mathematical treatment are substantial.

7.6 The Autocorrelation Function

For certain types of stochastic processes, the *autocorrelation function* is an important concept. Applications of this idea will be made in the next section.

Definition 7.2: Given a stochastic process $\{X_t\}$, the *autocorrelation function* for the process is the function of two variables R_X defined by

$$R_X(t_1, t_2) = E(X_{t_1} X_{t_2})$$

The subscript X in the expression R_X is, of course, the name of the stochastic process and is used simply to identify the stochastic process whose autocorrelation function is being discussed.

There are various situations in which the autocorrelation function for a random process is a useful tool. One which we will mention later has to do with what is called a "stationary" random process. Crudely speaking, a stationary process is one for which the choice of time to be considered as "time $t = 0$" is immaterial. One of the senses in which the term *stationary* is used involves the autocorrelation function for the process, and this idea will be featured in the next section. Meanwhile, we will analyze a few concrete examples.

Example 7.5. Consider the sine wave stochastic process

$$X_t = Y \cos \omega t$$

where ω is a constant and Y is a random variable which is uniform on the interval [0, 1]. This can be viewed as a sine wave in which the frequency and phase angle are fixed but in which the amplitude is given only in probabilistic terms. Each sample function will be simply a constant multiple (between 0 and 1) of the cosine function $\cos \omega t$. What are the expected value and the autocorrelation function of this stochastic process?

Solution:

$$E(X_t) = E(Y \cos \omega t) = E(Y) \cos \omega t = \frac{1}{2} \cos \omega t$$

$$R_X(t_1, t_2) = E(X_{t_1} X_{t_2}) = E(Y^2 \cos \omega t_1 \cos \omega t_2)$$

$$= \cos \omega t_1 \cos \omega t_2 \, E(Y^2) = \frac{1}{3} \cos \omega t_1 \cos \omega t_2$$

[That $E(Y^2) = 1/3$ was Problem 5.6.]

Example 7.6. Now let's consider a sine-wave process in which the amplitude and frequency are fixed but the phase angle is known only in a probabilistic sense. Specifically, we will consider $X_t = A \cos(\omega t + \theta)$, where A and ω are constants with $A > 0$ and where θ is a random variable uniform on the interval $[-\pi, \pi]$. What will the expected value and autocorrelation function of this stochastic process be?

Notice that in this example all the sample functions are simply translates of each other; that is, they all have the same amplitude and frequency and vary only in phase angle.

Solution: Each of the random variables X_t is now a function of the random variable θ. Specifically, $X_t = g(\theta)$ where g is the function of one variable $g(s) = A \cos(\omega t + s)$. This mean that we can use Proposition 3.7 and determine the expected value of X_t by using the density function for θ:

$$E(X_t) = \int_{-\pi}^{\pi} A \cos(\omega t + s) \frac{1}{2\pi} \, ds = \frac{A}{2\pi} \sin(\omega t + s) \Big]_{-\pi}^{\pi}$$

$$= \frac{A}{2\pi} (- \sin \omega t + \sin \omega t) = 0$$

Now for the autocorrelation function:

$$R_X(t_1, t_2) = E[A \cos (\omega t_1 + s) A \cos (\omega t_2 + s)]$$

$$= \int_{-\pi}^{\pi} A^2 \cos (\omega t_1 + s) \cos (\omega t_2 + s) \frac{1}{2\pi} ds$$

$$= \frac{A^2}{2\pi} \int_{-\pi}^{\pi} \frac{1}{2} [\cos (\omega t_1 + \omega t_2 + 2s) + \cos (\omega t_1 - \omega t_2)] ds$$

This last expression comes from the trigonometric identity

$$2 \cos x \cos y = \cos (x + y) + \cos (x - y)$$

The last integral is easily evaluated. Notice that one of the cosine terms does not contain s at all and therefore may be treated as a constant as far as integration with respect to s is concerned. The other term has a corresponding sine term as its antiderivative with respect to s, and since the endpoints of integration are $-\pi$ and π, the periodicity of the sine function makes that term integrate to 0. Thus, the answer is simply

$$R_X(t_1, t_2) = \frac{A^2}{2} \cos (\omega t_1 - \omega t_2)$$

7.7 Stationary Stochastic Processes

Keep in mind that in most common models, the parameter t in a stochastic process $\{X_t : t \geq 0\}$ represents time as measured on some discrete or continuous scale. So time $t = 0$ represents the "starting time" for the process. In many contexts, the starting time really is immaterial. For example, if you walk into a room where someone is tossing a coin, the probability that a head occurs on the next toss does not depend on how long the process has been going on before your arrival. It doesn't matter at what time you consider the process to have begun.

In a Poisson process, the random variables X_t and $X_{t_0+t} - X_{t_0}$ have the same distribution for any value of t_0. In terms of the phenomena being observed, this says that the number of occurrences during a time interval depends only on the length of the time interval and not on when the time interval begins. An even stronger condition is described in Definition 7.3.

Definition 7.3: A stochastic process $\{X_t\}$ is said to be a *stationary stochastic process* provided that for any values $t_1, t_2, \cdots, t_n$, and any $t > 0$, the joint probability distribution of the random variables $X_{t_1}, X_{t_2}, \cdots, X_{t_n}$ is the same as it would be if the subscripts $t_1, t_2, \cdots, t_n$ were replaced by $t_1 + t, t_2 + t, \cdots, t_n + t$. In other words, if all of the times were shifted by the same amount, this joint distribution of the random variables would remain the same.

Definition 7.3 treats a manner in which time shifts are irrelevant in some stochastic processes. This definition says, for instance, that the probability distribution for X_t is the same for every value of t. However, it says much more than this. Remember that the joint distribution contains all the information as to how the random variables interact with each other. The definition requires that all the details of this interaction be preserved if there is a time shift.

We can illustrate this definition by considering exactly what is and what is not stationary with regard to an independent trials process.

When one talks about an "independent trials process," there are really two stochastic processes lurking in the background. One is the process $\{X_n\}$, where X_n is the number of successes to occur during the first n trials. For each n, X_n is binomial with parameters n and p, where p is the success probability associated with the process. Clearly, if $n \neq m$ then X_n and X_m have different distributions, and so this stochastic process isn't stationary.

Rather than looking at the accumulated number of successes during the first n trials (which is what X_n tells us), we could consider the process $\{Y_n\}$, where Y_n is 1 or 0 depending upon whether success does or does not (respectively) occur on the nth trial. In other words, Y_n is the number of successes (0 or 1) that occur on the nth trial. This stochastic process $\{Y_n\}$ *is* stationary. This is an easy consequence of two facts: One is that $P(Y_n = 1) = p$ and $P(Y_n = 0) = 1 - p$ for every n (so the Y_n's all have the same probability distribution), and the other is that $Y_1, Y_2, \cdots$ are independent. The fact that they are independent means that the joint probability mass function of any finite collection of them is just the product of their individual mass functions which are all the same. The stochastic process $\{Y_n\}$ then clearly meets the criteria of Definition 7.3.

Standard terminology is to refer to $\{X_n\}$ as *the binomial process* and $\{Y_n\}$ as *the Bernoulli process*. Either can be defined in terms of the other quite easily:

$Y_1 = X_1$, and for $n > 1$, $Y_n = X_n - X_{n-1}$, and

$$X_n = \sum_{k=1}^{n} Y_k$$

In practice it is usually very difficult to show that a given stochastic process satisfies the conditions of Definition 7.3. The reason is that the joint distributions referred to may be very difficult or impossible to compute.

Stationary stochastic process have properties that are very useful in a variety of areas of application, such as signal processing. While Definition 7.3 is very difficult to check in practice, there is a weaker condition that is much easier to work with and is often used instead. It is introduced in Definition 7.4.

Definition 7.4: A stochastic process $\{X_t\}$ is said to be a *stationary in the wide sense* provided that

1. the expected value $E(X_t)$ remains the same for all values of the parameter t, and
2. the value of the autocorrelation function, $R_X(t_1, t_2)$, remains the same if t_1 and t_2 are both shifted by the same amount, that is,

$$R_X(t_1,t_2) = R_X(t_1 + t, t_2 + t) \quad \text{for every time shift } t$$

Notice that the computation shown in Example 7.6 was precisely the computation to show that the stochastic process $X_t = A \cos(\omega t + \theta)$, where A and ω are constant and θ is uniform on $[-\pi, \pi]$, is stationary in the wide sense. Stochastic processes that satisfy Definition 7.3 are often called *strictly stationary* to contrast them with the wide-sense stationary processes of Definition 7.4. It is easy to see that any strictly stationary stochastic process is also wide-sense stationary, but the converse isn't true. (See Problem 7.16.)

Proposition 7.4: If a stochastic process $\{X_t\}$ is strictly stationary, then it is also stationary in the wide sense.

Comment on Proof: From Definition 7.3, any two random variables X_{t_1} and X_{t_2} in the process must have the same probability distribution. It follows then that they must have the same expected value.

But why must $R_X(t_1, t_2) = R_X(t_1 + t, t_2 + t)$? We will illustrate in the case in which X_{t_1} and X_{t_2} are continuous random variables with a joint density function. This will enable us to view the computation in terms of Proposition 6.3.

Since $R_X(t_1, t_2) = E(X_{t_1} X_{t_2})$, it is first necessary to recall how this expected value can be computed in terms of the joint density function for X_{t_1} and X_{t_2}. The means for doing this is provided by Proposition 6.3:

$$E(X_{t_1} X_{t_2}) = \int_{-\infty}^{\infty}\int_{-\infty}^{\infty} xy\, f(x, y)\, dx\, dy \tag{7.4}$$

where f is the joint density function for X_{t_1} and X_{t_2}.

The significant point now is that X_{t_1+t} and X_{t_2+t} have exactly the same joint density function as do X_{t_1} and X_{t_2}. (This is what Definition 7.3 says.) So the expected value $E(X_{t_1+t} X_{t_2+t})$ would be described by exactly the same integral as the one that appears in Equation 7.4 above, and it is this that defines $R_X(t_1 + t, t_2 + t)$.

Example 7.7 (Random telegraph signal). Consider binary transmission of data sent along an electrical line. The two possibilities we will analyze are whether the voltage of the line is zero or nonzero relative to a ground voltage. Let's represent zero voltage by the value 0 and nonzero voltage by 1. At any time t, we can then think of the random variable X_t as indicating the state of the line at that time; that is, $X_t = 0$ if the voltage is zero and $X_t = 1$ if the voltage is nonzero at time t.

Furthermore, we will assume that there is some random "background phenomena" occurring which causes the line to toggle from one state to the other. An example could be something like cars entering a parking lot. As each car enters, it trips a switch which in turn sends a signal by changing the state of the transmission line being monitored. The situation we will consider is the case in which the background phenomena constitutes a Poisson process. So in the parking lot example, this means that for any given time interval the number of cars entering during the interval is a Poisson random variable with parameter $\lambda = \mu t$, where t is the length of the time interval and μ is the average number of cars entering the lot per unit time. Specifically, this leads to the following assumptions about the stochastic process $\{X_t\}$:

1. For any time t, $P(X_t = 0) = .5$ and $P(X_t = 1) = .5$.
2. During any time interval of length t, the probability that the line switches from the state $X_t = 0$ to the state $X_t = 1$ or vice versa exactly k times is given by

$$e^{-\mu t}\frac{(\mu t)^k}{k!}$$

The constant μ in condition 2 is the "intensity" of the underlying Poisson process, that is, the rate of the background phenomena that is causing the line to toggle back and forth between its two states. Condition 1 is a symmetry condition on the two possible states of the line. If the two voltage states are assumed equally likely to start with, the symmetry of the process suggests that the two states will always be equally likely.

Let's try to show that this "telegraphic" process $\{X_t\}$ is stationary in the wide sense.

First, for any t,

$$E(X_t) = 0 \times P(X_t = 0) + 1 \times P(X_t = 1) = .5$$

according to assumption (1).

Next we have to determine what the autocorrelation function $R_X(t_1, t_2)$ is for the process. Since $X_{t_1}X_{t_2}$ can only assume values 0 or 1,

$$\begin{aligned} E(X_{t_1}X_{t_2}) &= 0 \times P(X_{t_1}X_{t_2} = 0) + 1 \times P(X_{t_1}X_{t_2} = 1) \\ &= P(X_{t_1} = 1, X_{t_2} = 1) \end{aligned}$$

The reason for the last step is simply that the only way that the product of X_{t_1} and X_{t_2} can equal 1 is for each of them to be 1.

For definiteness, let's suppose that $t_1 < t_2$. The best way then to determine the probability $P(X_{t_1} = 1, X_{t_2} = 1)$ is to use the idea of conditional probability: $P(B \cap A) = P(A)\,P(B \mid A)$. This gives

$$P(X_{t_1} = 1, X_{t_2} = 1) = P(X_{t_1} = 1)\,P(X_{t_2} = 1 \mid X_{t_1} = 1)$$

Think about the conditional probability on the right side of this equation. If we know that $X_{t_1} = 1$, then the only way that X_{t_2} can equal 1 is for the line to toggle an even number of times between times t_1 and t_2. Since assumption 2 of this example gives the probability for a given number of changes in the state of the line during a given time interval, we can write the probability of an *even* number of changes during a time interval of length t as

$$e^{-\mu t}\left(1 + \frac{(\mu t)^2}{2!} + \frac{(\mu t)^4}{4!} + \frac{(\mu t)^6}{6!} + \cdots\right)$$

Since $P(X_{t_1} = 1) = .5$ by assumption 1, we now know that

$$R_X(t_1,t_2) = E(X_{t_1}X_{t_2}) = \frac{1}{2} \times e^{-\mu t}\left(1 + \frac{(\mu t)^2}{2!} + \frac{(\mu t)^4}{4!} + \frac{(\mu t)^6}{6!} + \cdots\right)$$

where $t = t_2 - t_1$ is the length of the time interval between times t_1 and t_2.

The remaining question is how to evaluate the sum of this infinite series on the right. This is not difficult if we make use of the standard power series representation for the exponential function. The sum of this series is

$$\frac{1}{2}\,(e^{\mu t} + e^{-\mu t})$$

(See Problem 7.17.) This gives us

$$R_X(t_1, t_2) = \frac{1}{2}\,e^{-\mu t} \times \frac{1}{2}\,(e^{\mu t} + e^{-\mu t}) = \frac{1}{4}\,(1 + e^{-2\mu t}) = \frac{1}{4}\,(1 + e^{-2\mu(t_2 - t_1)})$$

Notice that we have shown that this stochastic process *is* stationary in the wide-sense. This is a consequence of the fact that $R_X(t_1,t_2)$ depends here only on the *difference* $t_2 - t_1$. Consequently, if t_1 and t_2 were translated by the same amount, there would be no change in $R_X(t_1, t_2)$.

7.8 Ergodic Properties

Example 7.8. Let's return once more to the simple illustrative stochastic process given by $X_t = A\cos(\omega t + \theta)$, where A and ω are assumed to be constants and θ is assumed to be a uniformly distributed random variable on the interval $[-\pi, \pi]$. From Example 7.6 we already know that $E(X_t) = 0$ for all t.

Let's now take a look at

$$\frac{1}{T}\int_0^T X_t \, dt$$

This is the *time average* of the process X_t for the time interval $[0, T]$. The value will depend on what "observation" of the process we are looking at. In other words, since $X_t = A \cos(\omega t + \theta)$, the value we get for the time average will depend on what value between $-\pi$ and π we use for the random variable θ. If we carry out this averaging treating θ for the time being as just an unspecified real number, we get

$$\begin{aligned}\frac{1}{T}\int_0^T X_t \; dt &= \frac{1}{T}\int_0^T A \cos(\omega t + \theta) \; dt \\ &= \frac{1}{\omega T}\int_\theta^{\omega T+\theta} A \cos z \; dz \quad \text{(by a change of variables } z = \omega t + \theta) \\ &= \frac{A}{\omega T} \sin z \,\Big]_\theta^{\omega T+\theta} \\ &= \frac{A}{\omega T} [\sin(\omega T + \theta) - \sin\theta]\end{aligned}$$

Observe that the time average obtained here does depend on the value that the random variable θ assumes. However, if we take the limit of this as $T \to \infty$, we get 0, which is the expected value of X_t. This is true no matter what value θ assumes. So in fact we have

$$E(X_t) = \lim_{T\to\infty} \frac{1}{T}\int_0^T X_t \; dt \tag{7.5}$$

A stochastic process that satisfies this condition is said to be *ergodic in the mean.*

Notice that in order for this condition to be met, a few curious things have to be happening. Normally one would expect $E(X_t)$ to be different for different values of t. However, the right side of this equation is not dependent on t since it is a time average over the whole positive real line. Thus $E(X_t)$ must necessarily be the same for all values of t when Equation 7.5 is valid. Secondly, the right side would normally be expected to be dependent upon what "observation" of the process we are averaging, that is, what sample function. In this simple example, this means that we would normally expect the right size of Equation 7.5 to be different for different values assumed by the random variable θ. What we saw in the above computation is that this is, in fact, true for any finite interval. But when we average over the entire positive real line, such differences wash out.

In theory, the two kinds of averaging being performed on the two sides of Equation 7.5 are very different. The left side averages the value of a fixed random variable X_t over the sample space, and the right side averages all X_t's over time but corresponding to a fixed "outcome" in the sample space. There is an intuitive idea lurking behind all this. It is that over a long period of time the "average" behavior of the process is quite predictable.

Certainly not all stochastic processes have this ergodic property. Moreover, there are other forms of *ergodicity* that are of similar interest, such as ergodicity of the variance or of the autocorrelation function, for example. One characteristic that some models possess that leads to ergodic properties is that the behavior of the process during a given time interval may be "almost" independent of the behavior of the process during a distantly removed time interval. Assumptions of ergodicity frequently come into play when constructing models for real-world phenomena in which limited data is available. In modeling climatic phenomena, for example, one might have only one 50-year record of observations for a particular location. A reasonable assumption could be, however, that long-term cause and effect is negligible and therefore some future 50-year record may have similar statistics but be approximately independent from the existing record.

Problems

7.1 Use the Poisson distribution to approximate the probability of exactly 3 successes in 100 trials in an independent trials process with success probability $p = .04$.

7.2 Assume incoming calls at a switchboard to be a Poisson process with average intensity $\mu = .5$ calls per minute. What is the probability that no more than 2 calls come during a 2-minute time interval? What is the probability that no more than 2 calls come during a 3-minute time interval?

7.3 For an independent trials process with success probability p, derive the probability mass function for the random variable X that gives the number of the trial on which the *second* success occurs.

7.4 Suppose X and Y are independent geometrically distributed random variables, each with parameter p. Derive the probability mass function for $Z = X + Y$. Can you relate this exercise to Problem 7.3?

7.5 Using Equations 7.1 and 7.3, derive the probability density function for the random variable that measures the "waiting time until the second occurrence" in a Poisson process.

7.6 Suppose X and Y are independent exponentially distributed random variables, each with parameter λ. Derive the density function for the random variable $Z = X + Y$. Can you relate this to Problem 7.5?

7.7 Assume that breakdowns of a particular device constitute a Poisson process with an intensity of 10 breakdowns per year. What is the probability of getting 7 or more breakdowns during a 6-month period?

7.8 Calls are coming into a switchboard at a rate of 30 calls per hour. Assume this to be a Poisson process, and let X be the random variable that gives the number of incoming calls between 1:30 and 1:40, and Y the random variable that gives the number of incoming calls between 1:40 and 1:50. What kind of probability distribution does X have? How about Y? How about their sum $X + Y$? [Hint: Use the properties of Poisson processes described in this chapter. It would be possible to compute the mass function from basic principles based upon knowledge of the mass function of X and of Y and the fact that they are independent. In a Poisson process, however, the number of occurrences during *any* interval is a Poisson random variable where the parameter in the distribution is determined by the length of the time interval and the intensity of the process.].

7.9 Assume that automobile accidents in a city constitute a Poisson process with intensity $\mu = 5$ accidents per day. Compute the probability of only one accident occurring during a 1-day period, and compare this with the probability of two accidents occurring during a 2-day period. Do you intuitively understand why these probabilities are so different in magnitude?

7.10 Consider the sine-wave process $X_t = A \cos(\omega t + \theta)$, where ω is a constant and where A and θ are independent random variables with A uniform on $[0, 1]$ and θ uniform on $[0, 2\pi]$. Find the expected value of X_t as a function of t. (The function $t \to E(X_t)$ can be viewed as the average value of the process as a function of t.)

7.11 Let $Y_t = A \cos \omega t + B \sin \omega t$, where A and B both are uniform on $[0,1]$ and are independent and where ω is a fixed constant. Find $E(Y_t)$.

Recall that every function of the form

$$g(t) = a \cos \omega t + b \sin \omega t$$

where a, b, and ω are constants, can be written in the form

$$g(t) = c \cos (\omega t + \delta)$$

for the proper choice of c and δ. In fact,

$$c = \sqrt{a^2 + b^2}$$

and c is the amplitude of the wave.

For the stochastic process $Y_t = A \cos \omega t + B \sin \omega t$, this means that the amplitude is

$$Z = \sqrt{A^2 + B^2}$$

Notice that this is a random variable that does not depend on t. The expected value of this random variable, $E(Z)$, is then the expected amplitude. Write down an integral that would give the value of $E(Z)$. (Don't try to evaluate the integral.)

7.12 Let $X_t = A + Bt$, where A and B are independent and where both A and B have mean 1 and variance 1. Find the expected value and the variance of the stochastic process X_t. In other words, express $E(X_t)$ and $\text{var}(X_t)$ as functions of t.

7.13 Find $E(X_t)$ and $\text{var}(X_t)$, where $X_t = A \cos t$ and where A is standard normal.

7.14 A quantity of radioactive substance is emitting alpha particles at a rate of 2 particles per microsecond. Assume that this constitutes a Poisson process.

(a) What is the probability that more than 4 particles will be emitted during a fixed time interval of 2 μs?

(b) What is the probability that more than 6 particles will be emitted during a fixed time interval of 3 μs?

(c) If X is the number of particles emitted during a 1-μs time interval, and Y is the number emitted during a nonoverlapping 2-μs time interval, what is the probability that $X + Y$ will be greater than 6? Relate this to part b.

7.15 Suppose that Y is a non-negative continuous random variable. (This means that Y has a density function and that the density function is identically 0 on the negative part of the real line.) Consider the stochastic process

$$X_t = e^{-tY} \text{ for } t \geq 0$$

What is the relation between $E(X_t)$ and the Laplace transform?

7.16 Suppose $X_1, X_2, \cdots$ is a stochastic process as follows:

(1) When k is even, X_k assumes value 2 or −2, each with probability 1/2.

(2) When k is odd, X_k assumes value 1 with probability 4/5 and value −4 with probability 1/5.

(3) The random variables $X_1, X_2, \cdots$ are independent.

Show that this stochastic process is stationary in the wide sense but not in the strict sense. [Hint: It is easy to show that $E(X_k) = 0$ for all k and that $R_X(j, k) = 0$ if $j \neq k$ and $R_X(j, k) = 4$ if $j = k$. This is enough to establish that the process is stationary in the wide sense. On the other hand, for the process to be stationary would require that each of the random variables $X_1, X_2, \cdots$ have the same distribution. (This is the specific case that $n = 1$ in Definition 7.3.)]

7.17 Show that

$$e^{-\mu t}\left(1 + \frac{(\mu t)^2}{2!} + \frac{(\mu t)^4}{4!} + \frac{(\mu t)^6}{6!} + \cdots\right) = \frac{1}{2}\,(e^{\mu t} + e^{-\mu t})$$

This fact is used in Example 7.7.

[Hint: Let $f(x)$ be the sum of the series

$$f(x) = 1 + \frac{x^2}{2!} + \frac{x^4}{4!} + \cdots$$

Differentiate the series term by term, and show that

$$f(x) + f'(x) = e^x$$

Solve this differential equation to show that $f(x) = (1/2)\,(e^x + e^{-x})$.]

7.18 Flow of traffic into a parking lot has been cited as an example of a Poisson process. If the flow of traffic into the lot is regulated by a stop light, why then will the entrance of cars into the lot *not* be a Poisson process? [Hint: Which of the properties listed in Proposition 7.3 are not now satisfied?]

Chapter 8: Time-Dependent System Reliability

It is often assumed that "waiting time for a device to fail" is an exponentially distributed random variable. Chapter 7 sheds light on this in the context of Poisson processes. If the distributions of such "waiting time" random variables are assumed to be known, then the reliability analysis of simple systems as performed in Chapter 2 can be put in a time-dependent framework. In other words, probability calculations can give not just a raw number as the absolute reliability of a system; the reliability can instead be expressed as a function of time.

8.1 Reliability That Varies with Time

In Chapter 2 we talked about system reliability as a static concept. Each component of a system is assumed to have a known probability of failure, and the probability of system failure is then determined using these probabilities as inputs. Clearly one can envision situations when the reliability of components varies with time. New components may have a high failure rate because of "burn in" failures, and old ones may simply wear out. Engineers like to frame this discussion in terms of a function called the *hazard function* or *hazard rate* of a component. Let's relate this concept to some things that we already know.

Suppose that for a particular device, the time (from some reference time $t = 0$) until the device fails is a continuous random variable X. So X has a density function f_X and a distribution function F_X. The *reliability function* for the device is the function

$$R(t) = 1 - F_X(t) = P(X > t) \tag{8.1}$$

This means that $R(t)$ is simply the probability that the device is still functioning at time t. Since $F_X(t) \to 1$ as $t \to \infty$, it follows that $R(t) \to 0$ as $t \to \infty$.

The hazard function $h(t)$ for the device is defined then by

$$h(t) = \frac{f_X(t)}{1 - F_X(t)} = \frac{f_X(t)}{R(t)} \tag{8.2}$$

To see why this is a useful concept, notice first that the conditional probability $P(t \le X \le t + \Delta t \mid X > t)$ is the probability that the device fails between time t and time $t + \Delta t$, given that the device is still functioning at time t. The hazard rate $h(t)$ is

$$h(t) = \lim_{\Delta t \to 0+} \frac{P(t \le X \le t + \Delta t \mid X > t)}{\Delta t}$$

The reason for this is that

$$f_X(t) = F_X'(t) = \lim_{\Delta t \to 0+} \frac{F_X(t + \Delta t) - F_X(t)}{\Delta t} = \lim_{\Delta t \to 0+} \frac{P(t \le X \le t + \Delta t)}{\Delta t}$$

Thus,

$$\frac{P(t \le X \le t + \Delta t \mid X > t)}{\Delta t} = \frac{P(t \le X \le t + \Delta t)}{\Delta t \, P(X > t)}$$

$$= \frac{1}{1 - P(X \le t)} \times \frac{P(t \le X \le t + \Delta t)}{\Delta t}$$

$$\to \frac{f_X(t)}{1 - F_X(t)} \quad \text{as } \Delta t \to 0$$

Equation 8.2 describes the hazard rate $h(t)$ in such a way that it is easily computed if the density function of X is specified. It is also possible to turn the situation around and describe the density function or distribution function in terms of the hazard rate. The key relationship is

$$R(t) = 1 - F_X(t) = \exp\left\{ -\int_0^t h(s)\, ds \right\} \tag{8.3}$$

where exp denotes the exponential function. The derivation of this expression is as follows. If we integrate Equation 8.2, we have

$$\int_0^t h(s)\, ds = \int_0^t \frac{f_X(s)}{1 - F_X(s)}\, ds = -\ln\left[1 - F_X(s)\right]\Big|_0^t = -\ln[1 - F_X(t)]$$

The last step here assumes that $F_X(0) = 0$; that is, that at time $t = 0$ we know the device is in working condition. (Remember that f_X is the derivative of F_X. That's what the integration here is based on.) Applying the exponential function to both sides of this equation gives Equation 8.3.

A common scenario with real-world devices is a hazard rate something like what is shown in Figure 8.1. The curve is relatively high initially because of potential failure during the break-in period, and the curve rises again as the device ages. Clearly, not all devices behave this way, however. The most intuitively pleasant and easily understood situation is that of a constant failure rate; that is, $h(t)$ is a constant function. In this case Equation 8.3 becomes simply

$$R(t) = 1 - F_X(t) = e^{-\lambda t}$$

which is the familiar case of the exponential distribution. This is important because it is the canonical example for this concept.

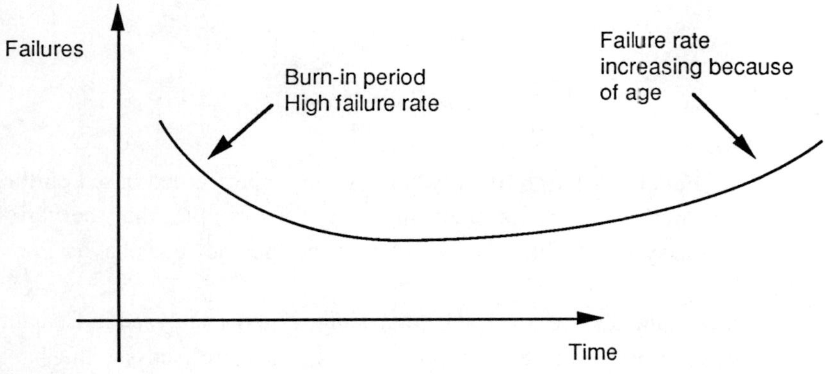

Figure 8.1 Graph of a typical hazard function.

Summary: To say that the failure rate of a device as given by Equation 8.1 is constant is precisely the same as saying that the "time to failure" for the device is exponentially distributed, that is, that the random variable that measures elapsed time until the device fails has an exponential density function.

One widely used statistic that is an indicator of the reliability of a device is the *mean time to failure (mttf)*. Intuitively, this is just the average waiting time from the time that the device is put into service until it first experiences failure. If X is the random variable that gives the time at which first failure occurs, then the mean time to failure for the device is simply $E(X)$.

Since the reliability of a device is often specified in terms of the reliability function, however, it is useful to be able to determine the mean time to failure in terms of the reliability function. Proposition 8.1 indicates how that is done.

Proposition 8.1: Suppose a device has reliability function $R(t)$. The *mean time to failure*, or *mttf*, for the device is given by

$$\text{mttf} = \int_0^\infty R(t)\,dt$$

Proof: If X is the random variable that indicates time to failure for the device, then by definition mttf $= E(X)$. However,

$$\int_0^\infty R(t)\,dt = \int_0^\infty 1 - F_X(t)\,dt = \int_0^\infty [1 - \int_0^t f_X(s)\,ds]\,dt$$

$$= \int_0^\infty \int_t^\infty f_X(s)\,ds\,dt = \int_0^\infty \int_0^s f_X(s)\,dt\,ds = \int_0^\infty s\,f_X(s)\,ds = E(X)$$

Time-dependent system reliability studies can be performed based on these concepts. Assumptions for the system might be, for example, that the various components of the system function independently and that they each have a known hazard function.

For a few examples, we will re-examine some systems analyzed in Chapter 2 with these ideas in mind. Figure 8.2 shows circuits that were discussed in Chapter 2. Now, however, we will assume that each component has a lifetime that is known to be an exponentially distributed random variable; that means that each component has a constant failure rate. For the components A, B, C, and D in the figure, we will denote by ρ_1, ρ_2, ρ_3, and ρ_4 the parameters in the exponential distribution for the lifetime of the respective devices; that is, the time to failure for device A is assumed to be an exponential random variable with parameter $\lambda = \rho_1$, and so on. (A bit more insight into understanding why this constant ρ_1 might be called the failure rate comes from remembering that if a random variable has exponential density with parameter λ, then the expected value is $1/\lambda$. Thus the expected value for the lifetime of component A is $1/\rho_1$, and so "on the average" that would correspond to ρ_1 failures per unit time. For example, if $\rho_1 = 2$ and time is measured in years, then the expected value for time until failure would be 1/2 year, which corresponds to a failure rate of two failures per year.)

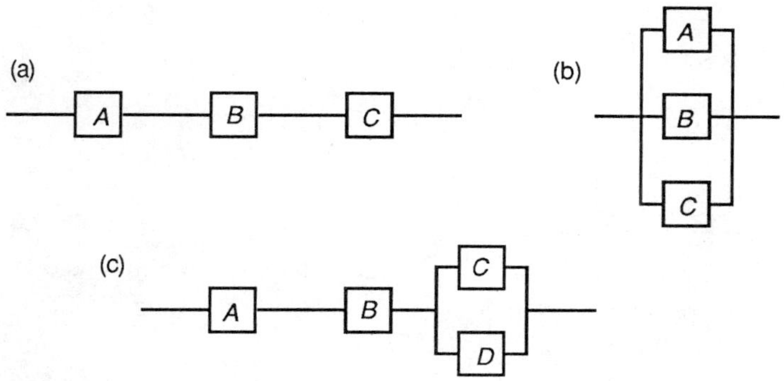

Figure 8.2 Three circuits.

In the circuits of Figure 8.2, let's denote by X, Y, Z, and W the times to failure of the components A, B, C, and D, respectively. So the assumption is that X, Y, Z, and W are exponential with parameters ρ_1, ρ_2, ρ_3, and ρ_4, respectively. Furthermore, we will assume that the lifetimes are independent; that is, X, Y, Z, and W form a set of independent random variables.

Since it is the system rather than an individual component that we are now considering, let's denote by $R_{\text{sys}}(t)$ the reliability function for the system. In other words, $R_{\text{sys}}(t)$ is the probability that the system is still functioning at time t.

For the series circuit in part a of the picture,

$$\begin{aligned} R_{\text{sys}}(t) &= P(X > t, Y > t, Z > t) = P(X > t)\, P(Y > t)\, P(Z > t) \\ &= e^{-\rho_1 t} e^{-\rho_2 t} e^{-\rho_3 t} = e^{-(\rho_1 + \rho_2 + \rho_3)t} \end{aligned}$$

From the form of this expression, we know that the "waiting time for the circuit to fail" is itself an exponential random variable with parameter $\lambda = \rho_1 + \rho_2 + \rho_3$. (The reason is simply that this reliability expression is equivalent to the statement that the distribution function for "time to failure for the circuit" is given by $1 - e^{-\lambda t}$, where $\lambda = \rho_1 + \rho_2 + \rho_3$.)

For the parallel circuit in part b of the picture, at least one device must still be functioning in order for current to flow, and so

$$\begin{aligned} R_{\text{sys}}(t) &= P\{ (X > t) \cup (Y > t) \cup (Z > t) \} \\ &= 1 - P(X \le t, Y \le t, Z \le t) \\ &= 1 - P(X \le t)\, P(Y \le t)\, P(Z \le t) \\ &= 1 - (1 - e^{-\rho_1 t})(1 - e^{-\rho_2 t})(1 - e^{-\rho_3 t}) \end{aligned}$$

For the series-parallel circuit in part c of the picture,

$$\begin{aligned} R_{\text{sys}}(t) &= P\{ (X > t) \cap (Y > t) \cap [(Z > t) \cup (W > t)] \} \\ &= P(X > t)\, P(Y > t)\, [1 - P(Z \le t)\, P(W \le t)] \\ &= e^{-\rho_1 t} e^{-\rho_2 t} [1 - (1 - e^{-\rho_3 t})(1 - e^{-\rho_4 t})] \end{aligned} \tag{8.4}$$

What are the probability densities of the time to failure for each of the three systems? For the series circuit shown in part a, the time to failure is exponentially distributed with parameter $\lambda = \rho_1 + \rho_2 + \rho_3$. So we know from Problem 3.15 that the expected time to failure (mean time to failure) is given by

$$\text{mttf} = \frac{1}{\rho_1 + \rho_2 + \rho_3}$$

For the parallel circuit shown in part b, the probability that the circuit will have failed by time t is

$$1 - R_{\text{sys}}(t) = (1 - e^{-\rho_1 t})(1 - e^{-\rho_2 t})(1 - e^{-\rho_3 t})$$

Since this is the distribution function for the random variable that represents waiting time for system failure, we can differentiate to get the density function

$$\begin{aligned} f(t) = {} & \rho_1 e^{-\rho_1 t} + \rho_2 e^{-\rho_2 t} + \rho_3 e^{-\rho_3 t} \\ & - \rho_1\rho_2 e^{-\rho_1\rho_2 t} - \rho_1\rho_3 e^{-\rho_1\rho_3 t} \\ & - \rho_2\rho_3 e^{-\rho_2\rho_3 t} + \rho_1\rho_2\rho_3 e^{-\rho_1\rho_2\rho_3 t} \end{aligned}$$

The mean time to failure is easy to compute here, because each of the exponential terms is of the form $\lambda e^{-\lambda t}$, and

$$\int_0^\infty \lambda t e^{-\lambda t}\, dt = \frac{1}{\lambda} \quad \text{for any } \lambda > 0$$

This implies that the mean time to failure for the parallel system is given by

$$\text{mttf} = \frac{1}{\rho_1} + \frac{1}{\rho_2} + \frac{1}{\rho_3} - \frac{1}{\rho_1\rho_2} - \frac{1}{\rho_1\rho_3} - \frac{1}{\rho_2\rho_3} + \frac{1}{\rho_1\rho_2\rho_3}$$

The density function for waiting time to system failure for the system shown in part c of the figure can be computed similarly. (See Problem 8.1.)

The computations taking place in these examples may appear to be quite a bit more complicated than the time-independent problems treated in Chapter 2. Let's try to understand precisely what the difference is in the two situations. The problems in Chapter 2 may be solved in a simple fashion because we can start putting numbers into the calculations whenever we want to. In other words, there is no necessity to work out a general reliability expression, a "formula" you might call it, into which we can substitute numbers corresponding to the reliabilities of the components and get the reliability of the system. However, reliability expressions have been derived in some examples. For instance, see Example 2.3. In principle, this can be done for more complex problems as well. Once the reliability expression is in hand, then we can make use of the fact that for each particular component the reliability function is known. Substituting these functions into the reliability expression gives the reliability function for the entire system.

To illustrate this concept, let's reconsider part c of Figure 8.2. Here the probability that this circuit is good is

$$P(A \text{ good})\, P(B \text{ good})\, P(C \text{ good or } D \text{ good}) = p_A p_B (p_C + p_D - p_C p_D)$$

Notice that in this expression it doesn't matter whether we are thinking of the reliability as being time-dependent or not. Now if we utilize the assumption that device A has an exponentially distributed lifetime with parameter ρ_1, this means that at time t the probability that A is good is $e^{-\rho_1 t}$. If we substitute $e^{-\rho_1 t}$ for p_A in the equation above and do a similar replacement for the other probabilities, the reliability function for the system is obtained just as in Equation 8.4. As a more elaborate example, see Problem 8.3. A solution using these ideas is given in Appendix A.

8.2 Systems with Repair

How can we model a device that is subject to random failures but in which there is also the capability of repairing the device after a failure occurs? After the device is repaired it can be placed back in service.

Let's assume a constant failure rate λ. This means that whenever the device is functioning, the waiting time for the next failure is assumed to be exponentially distributed with parameter λ. (Remember that the lack-of-memory property of the exponential distribution means that the time at which we begin measuring is irrelevant.)

We will similarly assume that the "time to repair" for the device is also exponentially distributed; that is, the total time that the device is out of service because of the failure is also an exponential random variable. We will denote by μ the parameter in the exponential distribution for the time the unit is out of service.

It is now possible to track the performance of the device over a long period of time. Presumably there would be a number of times that the device fails, and in each case there would be a certain delay in getting it serviced and put back into operation. What are some of the things that would be useful for us to know in this situation? One thing we would be interested in is the percentage of the time that the device is operational. For example, if breakdowns are so difficult to repair that the device is unavailable for use 60% of the time, this is certainly something that a prospective buyer would be interested in knowing. Or we might be interested in knowing the probability that the device will be functioning at some specific time in the future. This question is related to the former one. For instance, if we knew that the device was going to be out of service for 60% of the time during the next several years, that certainly suggests that over the long haul its probability of being operational ought to "average out" in some sense to 40%. Whether its probability of being operational next Tuesday afternoon is 40% is another question, and this suggests that we look at the situation in more detail.

First, let's remember some things we have already encountered. Since time to failure is exponentially distributed with parameter λ, we know that the mean time to failure for the device is $1/\lambda$. Similarly, since the time to repair is exponential with parameter μ, the mean time required for the device to be repaired is $1/\mu$. For example, if time is measured in years and $\lambda = .5$ and $\mu = 12$, then the mean time to failure will be 2 years and the mean repair time 1 month.

Now let's consider how we might actually work out the probability that the device will be functional at some specific time in the future. In order to do that, we need first to establish the necessary notation. Specifically, we would like to know about the following two functions:

$$f(t) = \text{probability the device is operational at time } t$$

$$g(t) = 1 - f(t) = \text{probability the device is } \textit{not} \text{ operational at time } t$$

We need to recall a few properties of the exponential function. If we think of the random variable X as representing time to failure for the device (as measured from initial time when the device is known to be working), then $P(X > \Delta t) = e^{-\lambda \Delta t}$. Now remember from the power series representation for the exponential function that

$$e^{-\lambda\Delta t} = 1 - \lambda\Delta t + \frac{(\lambda\Delta t)^2}{2!} + \cdots$$

In particular, if Δt is small, then $e^{-\lambda\Delta t} \approx 1 - \lambda\Delta t$. (This approximation amounts to using a linear approximation to evaluate the exponential function near 0.) This means that

$$P(X > \Delta t) \approx 1 - \lambda\Delta t$$

when Δt is small, and it means that the probability that a failure *will not* occur before time Δt is approximately $1 - \lambda\Delta t$. Conversely, the probability that a failure *will* occur before time Δt is approximately $\lambda\Delta t$. Applying this same idea to the time to repair leads to the conclusion that for a small time interval Δt, if we know the device to be nonoperational at the beginning of the time interval, then the probability that it will still be nonoperational after time Δt has elapsed is approximately $1 - \mu\Delta t$, whereas the probability that it will have been repaired during the time interval is approximately $\mu\Delta t$.

We now need to consider $f(t + \Delta t)$, where t is some fixed time and Δt is considered a small positive number. By definition, $f(t + \Delta t)$ is the probability that the device is working at time $t + \Delta t$.

There are two primary ways that the device can be working at time $t + \Delta t$: (1) It could be working at time t and experience no failure between time t and time $t + \Delta t$. (2) It could have been out of service at time t but have been repaired between time t and $t + \Delta t$. There are clearly other possibilities involving more than one failure and/or repair during the time interval, but if Δt is small then the probabilities of such terms will be small compared to the terms above. (If all of this seems a little fishy, it might be a good idea to go back and review the discussion of Poisson processes in Chapter 7. The assumptions here are quite similar to those that describe the Poisson process.)

So we can break $f(t + \Delta t)$ down as follows:

$$\begin{aligned} f(t + \Delta t) \approx\ & \text{probability device works at time } t \text{ and} \\ & \qquad \text{doesn't fail during time interval of length } \Delta t \\ & + \text{probability device is failed at time } t \\ & \qquad \text{but is repaired during interval of length } \Delta t \\ \approx\ & f(\text{t})\,(1 - \lambda\Delta t) + g(t)\,\mu\Delta t \end{aligned}$$

If we rearrange the terms and replace $g(t)$ by $1 - f(t)$, we can write this as

$$\frac{f(t+\Delta t) - f(t)}{\Delta t} \approx -\lambda f(t) + \mu g(t) = -(\lambda + \mu)f(t) + \mu$$

Taking the limit as $\Delta t \to 0$ leads to the differential equation

$$f'(t) = -(\lambda + \mu)f(t) + \mu$$

This is a simple first-order linear equation and has the solution

$$f(t) = ce^{-(\lambda+\mu)t} + \frac{\mu}{\lambda+\mu} \tag{8.5}$$

The value of the constant c will of course be determined by the initial conditions. For instance, if we know for certain that the device is working at time 0, then $f(0) = 1$, and this leads to the solution

$$f(t) = (1 - \frac{\mu}{\lambda+\mu})\, e^{-(\lambda+\mu)t} + \frac{\mu}{\lambda+\mu} \tag{8.6}$$

for the probability that the device is working at time t. Moreover, since $g(t) = 1 - f(t)$, the probability that the device is *not* working at time t is

$$g(t) = \frac{\lambda}{\lambda+\mu}\,(1 - e^{-(\lambda+\mu)t}) \tag{8.7}$$

The situation is actually more symmetric than the last two equations make it appear. The asymmetry stems from the fact that $f(0) = 1$ whereas $g(0) = 0$. The general solution for g is the same as the general solution for f except that the roles of μ and λ are reversed.

Notice that as time passes the effect of the initial conditions dies out.

$$f(t) \to \frac{\mu}{\lambda+\mu} \quad \text{and} \quad g(t) \to \frac{\lambda}{\lambda+\mu} \text{ as } t \to \infty$$

So if $\lambda = .5$ and $\mu = 12$ as in the earlier illustration, then as time passes the probability that the device will be working asymptotically approaches .96 and the probability that the device is not working approaches .04.

The model which we have just been using for a repairable device is a special case of an important class of stochastic processes called *Markov processes.* A Markov process is a stochastic process $\{X_t\}$ in which the random variable X_t describes the *state* of a system at time t. In the example we have just considered, there are two states: The device is *working* or it is *not working*. Clearly if there are a finite number of states, we may then think of them as being represented by integers $1, \cdots, n$. Thus the random variables that make up a Markov process may be viewed as taking values $1, \cdots, n$ for some integer n, with the values representing the different states of some system. The study of Markov processes revolves around consideration of the *transition probabilities* associated with moving from one state into another. In our simple example we could consider "working" as state 1 and "not working" as state 2. The length of time that we remain in each state was assumed exponentially distributed in our example, and in our case there is no doubt about which state we will move to when a change of state takes place. In a more general setting with numerous states involved, the transition probabilities specify this information. For example, if one is modeling the number of items waiting to be repaired at a service shop and if the items are being delivered and processed individually, then the "state" of the system would be the number of items present, and a change of state would be caused by (1) finishing work on an item and shipping it out or (2) a new item being brought to the shop. If the state is k, then a change of the first type makes the new state of the system $k - 1$, and a change of the second type makes the new state $k + 1$. Whether the transition probability of going from state k to state $k - 1$ is greater or less then the probability of going from state k to $k + 1$ is then simply a question of whether the items are being processed faster than they are arriving.

The precise definition of a Markov process is somewhat technical, but the general idea is that the probability of making a transition from one state (state i) to another (state j) in a certain time period should not be dependent on how one came to be in state i in the first place. In our problem above, this assumption is implicitly present in the assumed exponential waiting times for a change of state. From any point in time and from either of the two states, future transitions are governed by the current state of the device and by the two exponential distributions involved and are independent of the past history of the device.

Problems

8.1 Determine the mean time to failure (mttf) for the system in Figure 8.2(c) assuming that the components A, B, C, and D function independently and have constant failure rates ρ_1, ρ_2, ρ_3, and ρ_4 respectively, that is, assuming that the time to failure for each component is an exponential random variable with the given parameter.

8.2 Consider a device with hazard function given by $h(t) = 1 + (2 - t)^2$ whenever $t > 0$. If we think of time as measured in years, you can notice that the instantaneous failure rate is dropping off until the age of the device reaches 2 years, at which time the failure rate starts to increase again because of age. Let X denote "time to failure." Determine the density function for

X as well as the distribution function. (Actually, the latter should be done first, using Equation 8.3.) What is the probability that the device fails during its second year, that is, between time $t = 1$ and time $t = 2$? Notice that the mttf here is not easily computable because $tf_X(t)$ is not readily integrable.

8.3 In the circuits shown in Problem 2.2 of Chapter 2, assume that each component has an exponentially distributed lifetime with mean 1 year, and that the components function independently. Find the reliability function

$$R_{\text{sys}}(t) = \text{probability that the system is still functioning at time } t$$

8.4 In the highway network shown in Problem 2.4 of Chapter 2, assume that each link in the network (each edge of the graph) has an exponentially distributed lifetime with mean 6 weeks. Find the reliability function

$$R_{\text{sys}}(t) = \text{probability of being able to go from Start to Finish at time } t$$

8.5 Determine the density function for "waiting time to system failure" for the series-parallel system in the following figure.

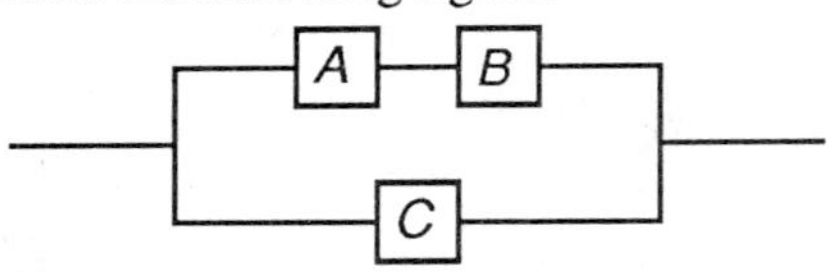

Assume that the components function independently, that each has an exponentially distributed time to failure, and that the mean time to failure for the individual components are 5 months for component A, 10 months for component B, and 20 months for component C. What is the mttf for the system?

8.6 Suppose a device has a hazard function that is a linear function of time; that is, $h(t) = kt$, where k is a positive constant. (This says that the rate at which such devices fail is proportional to the age of the devices.) Find the reliability function for such a device.

8.7 A piece of equipment averages going 90 days between breakdowns, and the time requirement to get it serviced averages 3 days. Answer the following questions using the model developed in Section 8.2.

(a) In the long run, what fraction of the time is the equipment operable?

(b) If it is operating now, what is the probability it will be operating 5 days from now? [Hint: Treat "now" as time zero and the given information that it is operating at time zero as an initial condition.]

8.8 A telephone is a two-state device in that the line is either free or busy. Assume that when the phone is not being used, the time until a shift into the "busy" state is exponentially distributed with expected value 10 minutes. And assume that when the phone is busy, the time until a shift into the *not-in-use* state is exponentially distributed with expected value 3 minutes. Based on the model of Section 8.2, answer the following:

(a) If the phone is busy at time $t = 0$, find the function $f(t)$ = probability that the phone will be in use at time t.

(b) In the long run, what fraction of the time is the phone in use?

8.9 (A modeling problem) A telephone has a "hold" button so that an incoming call can be placed on hold if the line is tied up. So there are three states for this phone: (1) the line is free, (2) the line is busy, and (3) the line is busy and someone is on hold.

Any incoming call that arrives while the phone is in state 3 is turned away. Thus from state 3 it is possible to go only to state 2, and similarly from state 1 it is possible to go only to state 2. From state 2 we might go to either state 1 or state 3, depending on whether and incoming call arrives before the present caller hangs up. The probabilities will depend on the relative rate of incoming calls as compared to the length of time that the calls last.

We will assume that arriving calls form a Poisson process with intensity λ and that the duration of calls is exponentially distributed with parameter μ. So the length of time in state 3 (the length of time until the present caller hangs up) is exponential with parameter μ, and the length of time in state 1 (the length of time until an incoming call) is exponentially distributed with parameter λ. (Review the properties of a Poisson process if you need to.) But what about the length of time in state 2? When in state 2, the phone will stay in state 2 until *either* the caller hangs up (sending us back to state 1) *or* another call arrives (sending us to state 3). Recall that the *minimum* of two exponential random variables with parameters μ and λ is exponential with parameter $\mu + \lambda$. So the waiting time for a change of state while in state 2 is exponentially distributed with parameter $\mu + \lambda$. When a change of state *does* occur from state 2, what is the probability it will be a change to state 1? This is simply the question: What is the probability that the present caller will hang up *before* the next incoming call arrives? We solved this problem in Chapter 6. (See Problem 6.26.)

If $f(t)$, $g(t)$, and $h(t)$ represent the probabilities that at time t we will be in states 1, 2, and 3, respectively, see if you can derive the system of differential equations that must be satisfied by these three functions. You can do so by mimicking the approach used in Section 8.2.

References

1. Breiman, Leo, *Probability and Stochastic Process with a View toward Applications,* Houghton Mifflin, Boston, 1969.

2. Chung, Kai Lai, *Probability Theory with Stochastic Processes,* Springer-Verlag, New York, 1974. A very readable general-purpose text.

3. Cinlar, Erhan, *Introduction to Stochastic Processes,* Prentice-Hall, Englewood Cliffs, N.J., 1975. This book presupposes a solid foundation in probability and advanced calculus. The exposition is aimed, however, at engineers and others who have applications rather than theory as their objective.

4. Dwass, Meyer, *Probability Theory and Applications,* W. A. Benjamin, New York, 1970. This book is a good reference on specific distributions. It includes treatments of the gamma and beta distributions, the chi-square distribution, and the multinomial and negative binomial distibutions. The exposition is a bit more theoretical than is found in engineering oriented texts.

5. Feller, William, *An Introduction to Probability Theory and Its Applications,* Wiley, New York, 1968. This is perhaps the most widely referenced work on probability in existence. It consists of two volumes, the first on discrete distributions, and the second on continuous distributions. These books include much information not found in general-purpose texts.

6. Page, Lavon B., and Jo Ellen Perry, "A practical implementation of the factoring theorem for network reliability," *IEEE Trans. Reliability,* R-36, pp. 259–267, 1988. This article describes a microcomputer algorithm which uses conditional probabilities to treat network reliability problems in a manner reflecting the ideas encountered in Chapter 2.

7. Solomon, Frederick, *Probability and Stochastic Processes,* Prentice-Hall, Inc., Englewood Cliffs, N.J., 1987. This is a general-purpose undergraduate probability text.

Appendices

Appendix A: Answers, Partial Solutions, and Hints to Selected Problems

Appendix B: Values of the Normal Distribution Function

Appendix A : Answers, Partial Solutions, and Hints to Selected Problems

Chapter 1

1.1 (a) .18 (b) .6 (c) .72 (d) .88

1.2 This problem can be done easily using a tree diagram or the combinations formula. If the combinations formula is used the answers can be given by

(a) $\dfrac{C(4, 2)}{C(6, 2)} = \dfrac{2}{5}$ (b) $\dfrac{2/5}{1 - C(2, 2)/C(6, 2)} = \dfrac{3}{7}$

1.3 .3

1.4 $\dfrac{C(4, 2) + C(3, 2) + C(2, 2)}{C(9, 2)} = \dfrac{5}{18}$

1.5 $P(A) = .4$

1.6 (a) $P(2 \text{ good}) = C(5, 2)\,(.9)^2(.1)^3 = .0081$, and so on.
(b) $P(\text{at least 3 good}) = P(3 \text{ good}) + P(4 \text{ good}) + P(5 \text{ good})$

1.8 The answers are easily obtained from following tree diagram.

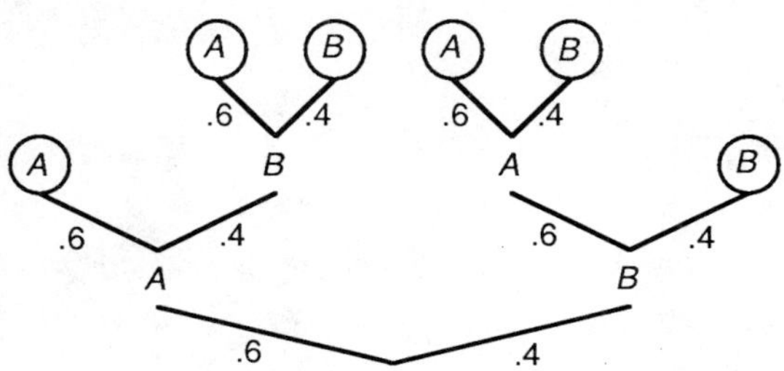

(b) $(.6)^2 + 2(.6)^2(.4)$ (c) $2(.6)^2(.4)$ (d) 64 (e).36

1.11 Using the following tree, we see that $P(\text{disease and } +) = .045$ and $P(\text{disease} \mid +) = .045/.121 = .3719$.

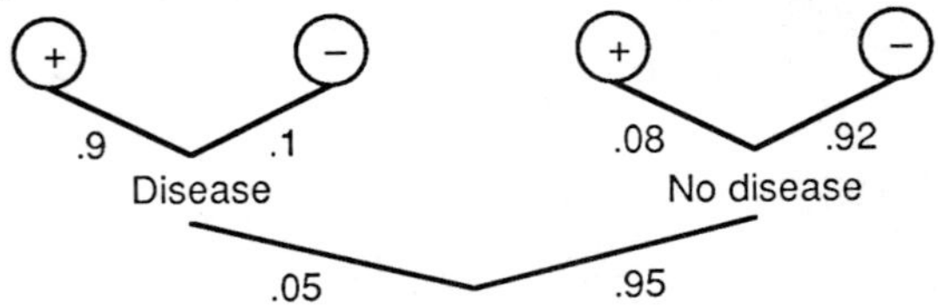

1.13 (a) $P(B) = P(A_1)\,P(B \mid A_1) + \cdots + P(A_6)\,P(B \mid A_6)$

$$= \frac{1}{6} \times \frac{5}{6} + \frac{1}{6} \times \frac{4}{6} + \cdots + \frac{1}{6} \times \frac{1}{6} = \frac{15}{36}$$

(b) A_3 is the event that the green die shows a 3. So

$$P(A_3 \mid B) = \frac{P(A_3)\,P(B \mid A_3)}{\sum_{i=1}^{6} P(A_i)P(B \mid A_i)} = \frac{1/6 \times 3/6}{15/36} = \frac{1}{5}$$

1.14 (a) $S = \{(\text{H}, 1), (\text{H}, 2), (\text{H}, 3), \cdots, (\text{T}, 5), (\text{T}, 6)\}$

(b) $P(\text{head on coin}) = 6/12$, $P(\text{even number on die}) = 6/12$, $P(\text{head on coin and even no. on die}) = 3/12$.

1.17 $C(10, 2)\,(.3)^2(.7)^8 = .233474$

1.18 $$\frac{C(30, 2)\,C(70, 8)}{C(100, 10)} = .237232$$

1.22 If n is the number of random digits generated, then $P(\text{no 7's}) = (.9)^n$. Therefore, $P(\text{at least one 7}) = 1 - (.9)^n$. So what we want is to have n sufficiently large that $1 - (.9)^n \geq .95$. This means we need $(.9)^n \leq .05$. Take the natural logarithm of both sides: To have $n \ln(.9) \leq \ln(.05)$ requires that $n \geq \ln(.05)/\ln(.9) \approx 28.43$. Thus we need to generate at least 29 random digits.

Chapter 2

2.2 (a) .58224 (b) .95086

2.4 Going from start to finish is possible provided that link E is good and that at least one of the two pairs A, B or C, D is good. The probability is

$$P\{\ [(A \cap B) \cup (C \cap D)] \cap E\ \}$$
$$= P[\ (A \cap B) \cup (C \cap D)\ ] \times P(E)$$
$$= [P(A \cap B) + P(C \cap D) - P(A \cap B \cap C \cap D)] \times P(E)$$
$$= p_A\, p_B\, p_E + p_C\, p_D\, p_E - p_A\, p_B\, p_C\, p_D\, p_E$$

2.8 .970299

2.9 $P(\text{battery}) \times P(R_1) \times P(\text{light}) \times P(R_2 \text{ or } R_3)$
$= P(\text{battery}) \times P(R_1) \times P(\text{light}) \times \{P(R_2) + P(R_3) - P(R_2) \times P(R_3)\}$

2.10 Suppose n of the devices are used. The probability they all fail is then $.2^n$, so the probability that at least one works is $1 - .2^n$. We wish this probability to be $\geq .9999$. So we need to have n large enough that $1 - .2^n \geq .9999$. This inequality is easily solved for n: The inequality $.2^n \leq .0001$ will be true whenever $n \ln(.2) \leq \ln(.0001)$. This requires that $n \geq 5.72$. Since a whole number of devices is required, we should use $n = 6$ devices.

Chapter 3

3.1 Your graph should have a jump discontinuity at 0, 1, 2, 3, and 4 and should be constant on any interval not containing one of these numbers.

3.3

$$F_X(t) = \begin{cases} 0 & \text{if } t < 0 \\ t & \text{if } 0 \leq t \leq 1 \\ 1 & \text{if } t > 1 \end{cases}$$

3.4 (a) $P(0 < X < .5) = .25$ (b) $F_X(.5) = .75$

(c) $$F_X(t) = \begin{cases} 0 & \text{if } t < -1 \\ \dfrac{t+1}{2} & \text{if } -1 \le t \le 1 \\ 1 & \text{if } t > 1 \end{cases}$$

3.5 $P(X = 1) = 1/36$, $P(X = 2) = 3/36$, $P(X = 3) = 5/36$, and so on.

3.6 This random variable could be interpreted as what you're looking at if you choose a random number between 0 and 2 and truncate it after the first decimal place.

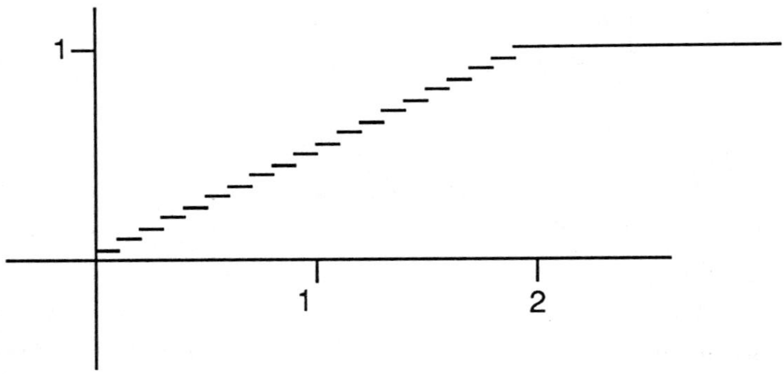

3.7 $P(X > 1) = e^{-3}$ and $P(1 < X < 2) = e^{-3} - e^{-6}$.

3.8 $P(|X| < 1) = P(-1 < X < 1) = 1 - e^{-1}$

3.9 $c = \dfrac{3}{32}$

3.12 Skip ahead and look at Definition 4.1. In Chapter 4 the random variable X will be called a binomial random variable with parameters $n = 4$ and $p = .1$.

3.13 $E(X) = 1/2$ and $E(Y) = 2$.

3.16 $F_Y(t) = P(Y \le t) = P(2X+3 \le t) = P[X \le (t-3)/2] = F_X[(t-3)/2]$. You computed F_X in Problem 3.3. Using that information we get

$$F_Y(t) = \begin{cases} 0 & \text{if } t < 3 \\ \dfrac{t-3}{2} & \text{if } 3 \le t \le 5 \\ 1 & \text{if } t > 5 \end{cases}$$

To obtain f_Y simply differentiate F_Y. What you will notice is that Y is uniformly distributed on the interval [3, 5]. This should not be surprising given the way that Y is defined in terms of X.

3.17 Notice that Problem 3.16 is a special case of this.

3.18 (c) $E(X) = .21$

Chapter 4

4.1 For example,

$$P(X=2) = \frac{C(4, 2)\,C(9, 1)}{C(13, 3)}$$

4.2 (a) $\dfrac{1}{2^5} = \dfrac{1}{32}$ (b) $C(10, 3)\,\dfrac{1}{2^{10}}$

4.3 (a) $E(X) = .01\displaystyle\sum_{k=1}^{100} k$ (c) $E(Y) = .01\displaystyle\sum_{k=1}^{100} 2^k$

4.4 X = number of heads in 3 tosses, Y = number of tails = $3 - X$. So the number of heads minus the number of tails is $X - (3 - X) = 2X - 3$. Therefore $E(Y)$ is given by

$$\sum_{k=0}^{3} |2k-3|\,P(X=k) = 1.5$$

4.6 Show that $\dfrac{P(X = k)}{P(X = k - 1)} = \dfrac{\lambda}{k}$

This shows that $P(X = k) > P(X = k - 1)$ if and only if $\lambda > k$. So if we look at the probabilities $P(X = 1), P(X = 2), \cdots$, the probabilities are getting larger until k gets to be bigger than λ.

Conclusion: The largest probability is $P(X=k)$ where k is the largest integer less than λ. If λ is itself an integer, then there is a tie between the two numbers $P(X = \lambda)$ and $P(X = \lambda + 1)$.

4.7 $P(X = 3) = .14037$

4.8 X is binomial with $n = 50, p = .1$, and $P(X = 3) = .13857$.

4.10 $\text{var}(X) = 2$

4.19 .2646

4.21 The distribution to use is the binomial distribution.

Chapter 5

5.2 (a) .6065 (b) .6321 (c) .6321

5.3 $Y = 1 - X$. This means that

$$\begin{aligned} F_Y(t) &= P(Y \le t) = P(1 - X \le t) = P(X > 1 - t) \\ &= 1 - P(X \le 1 - t) = 1 - F_X(1 - t) \end{aligned}$$

Use your knowledge of what F_X is (Problem 3.3) and be very careful with your inequalities to show that in fact $F_Y = F_X$, and so Y also is uniform on [0, 1].

5.4 (a) The answer is 1/2. You don't need a table if you just remember the symmetry of the density function.

(b) .36788

5.5 The method is similar to Exercise 5.3.

(c) $f_Y(t) = \begin{cases} \frac{1}{2} e^{-(t-1)/2} & \text{if } t > 1 \\ 0 & \text{otherwise} \end{cases}$ (d) $E(Y) = 3$

5.6 f_Y is given by

$$f_Y(t) = \begin{cases} \frac{1}{2\sqrt{t}} & \text{if } 0 < t < 1 \\ 0 & \text{otherwise} \end{cases}$$

$E(Y) = \frac{1}{3}$

5.7 $Y = |X - 1|$. This means

$$\begin{aligned} F_Y(t) &= P(Y \le t) \\ &= P(|X - 1| \le t) \\ &= P(-t \le X - 1 \le t) = P(1 - t \le X \le 1 + t) \end{aligned}$$

Since the distribution of X is known, this probability can be easily evaluated. You probably will want to treat the three cases $t < 0$, $0 \le t \le 1$, and $1 < t$ separately. What you should find out is that F_Y turns out to be uniform on [0, 1]. Think about the way Y is defined, and try to understand intuitively why this is true.

5.8 This problem is really a combination of two problems already done. Think of the wire as being laid on the x-axis from 0 to 2. If X is the x coordinate of the point where the cut is made, the assumption is that X is uniform on [0, 2]. Figure out that the length Y of the shortest piece is the random variable $Y = 1 - |X - 1|$. To put this in the context of problems already done, let's temporarily use the notation $Z = |X - 1|$. From Problem 5.7 we know that Z is uniform on [0, 1]. Since $Y = 1 - Z$, we know from Problem 5.3 that Y then is also uniform on [0, 1], and so $E(Y) = 1/2$. The expected length of the shortest piece of wire is 6 inches.

Of course it is not absolutely necessary to view this problem in terms of Problems 5.3 and 5.7. This problem could be done from scratch using the same techniques as are used in those problems.

5.9 6.93 years

5.11 $P(2 < X < 4) = \Phi(3/2) - \Phi(1/2) = .2417$, where Φ is the standard normal distribution function.

5.12 $\text{var}(X) = E(X^2) - E(X)^2$. Compute $E(X^2)$ as follows using integration by parts:

$$E(X^2) = \int_0^\infty t^2 \lambda e^{-\lambda t}\, dt$$

5.15 Let X = number of defects. Then

$$\begin{aligned} P(225 < X < 275) &= P\left(\frac{225-\mu}{\sigma} \le \frac{X-\mu}{\sigma} \le \frac{275-\mu}{\sigma}\right) \\ &= \Phi(-1/4) - \Phi(-3/4) \\ &= \Phi(3/4) - \Phi(1/4) = .1747 \end{aligned}$$

5.17 (a) .6065 (b) .3679 (c) .1353

5.18 (a) $.98^{25} \approx .6035$ (b) $.98^{50} \approx .3642$ (c) $.98^{100} \approx .1326$

5.22 Let X denote the random number, and let Y denoted the value of X truncated to a single digit after the decimal. Then $Y = g(X)$ where g is the step function shown in Figure 4.1. This means that $E(Y)$ can be computed as

$$E(Y) = \int_{-\infty}^{\infty} g(t) f_X(t)\, dt = \int_0^1 g(t)\, dt = .45$$

5.24 .15

5.25 The equation $E = IR$ is used here, where E = voltage drop, I = current, and R = resistance. Applying this equation to the entire circuit gives $12 = I(R+1)$. Also from this equation we know that the voltage drop V across the resistor is given by $V = IR$. So $I = 12/(R+1)$, and this implies that $V = 12R/(R+1)$. Therefore,

$$F_V(t) = P(V \le t) = P[\,12R/(R+1) \le t\,] = P[\,12R \le t(R+1)\,]$$

$$= P[\,(12-t)\,R \le t\,] = P[\,R \le t/(12-t)\,] = F_R[\,t/(12-t)\,]$$

The distribution function F_R is familiar. Figure out what it is, and from that you can deduce that

$$F_V(t) = \begin{cases} 0 & \text{if } t \le 6 \\ \dfrac{t}{12-t} & \text{if } 6 \le t \le 8 \\ 1 & \text{if } t \ge 8 \end{cases}$$

It is easy to obtain the density function for R just by differentiating F_V.

As usual, $E(V)$ can be computed from f_V or directly from the density function for R.

Chapter 6

6.1 The table should look something like this:

	0	1	*X*
0	1/4	0	**1/4**
1	1/4	1/4	**1/2**
2	0	1/4	**1/4**
Y	**1/2**	**1/2**	

6.3 The computations to obtain the table below are a bit tedious, but not terribly difficult if you get the probabilities from a tree diagram.

	0	1	2	*X*
0	1/15	4/15	1/15	**2/5**
1	2/15	4/15	2/15	**8/15**
2	0	1/15	0	**1/15**
Y	**1/5**	**3/5**	**1/5**	

6.5 This one is easy. The table is as follows:

	0	1	2	X
0	1/8	1/8	0	1/4
1	1/8	1/4	1/8	1/2
2	0	1/8	1/8	1/4
Y	1/4	1/2	1/4	

6.7 Partial answer: $E(X) = E(Y) = 2/5$.

6.12 Let X and Y be the two numbers. The first step is to realize that the probability in question can be obtained by integrating the joint density function $F_{X,Y}$ over the half-plane below the line $y = x/2$. Secondly, the joint density function is easily obtained from Definition 6.6 since X and Y are independent. The problem then boils down to integrating the joint density function over the shaded triangle in the picture. Since the joint density is simply the uniform density on the unit square, the probability is just the area of the triangle, which is 1/4.

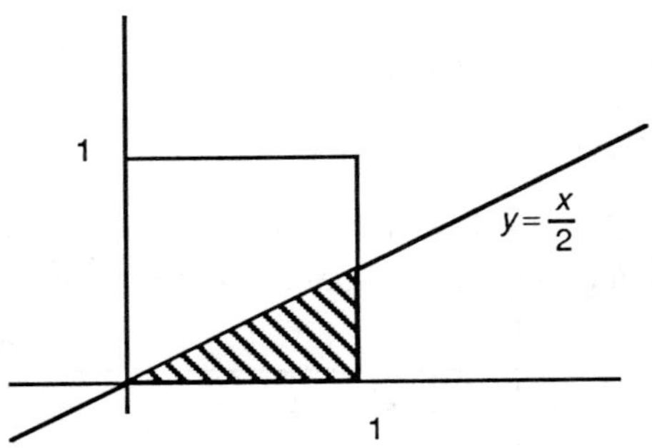

6.15 The density function for X is: $f_X(x) = 1/(1 + x)^2$ whenever $x > 0$, with $f_X(x) = 0$ when $x < 0$. The density for Y is simply the exponential density with parameter $\lambda = 1$.

6.17

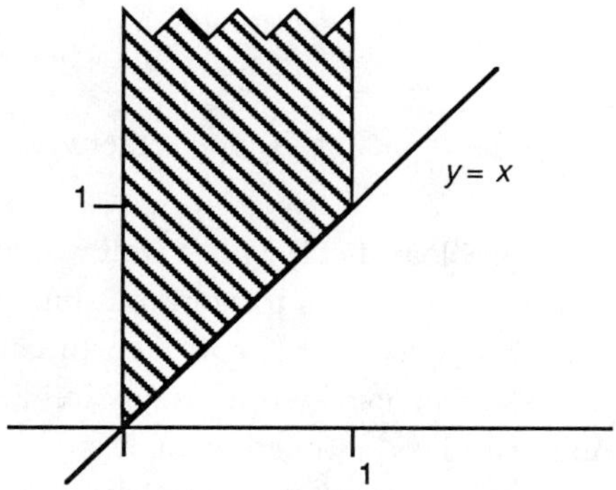

Set time $t = 0$ to be 8 o'clock. Let X be the time when your friend calls you, so the assumption is that X is uniform on [0, 1]. Let Y be the waiting time for the first call from someone *other* than your friend, so Y is exponential with parameter $\lambda = 2$. (Remember that in the exponential distribution the parameter is the reciprocal of the expected value.)

The problem is to find $P(X < Y)$. This is computed by integrating the joint density function over the region consisting of the half-plane above the line $y = x$. The joint density is 0 except when $0 < x < 1$ and $y > 0$, and the part of this region that lies in this half-plane is the region shaded in the picture. So the probability is computed by performing the integration

$$\int_0^1 \int_x^\infty 2e^{-2y}\, dy\, dx = .4323$$

6.19 $E(X + Y) = \dfrac{520}{3}$

6.20 (a) It is easy to show that both X and Y have exponential densities.

(b) Yes.

(c) $P(X > 10\,,\, Y > 10) = e^{-3}$.

(d) Integrate the joint density over the appropriate triangle and subtract from 1.

6.22 It should be apparent that this computation is going to require integrating the joint density over a triangle in the first quadrant. The density function works out to be like this: If $Z = X + Y$, then

$$f_Z(t) = \begin{cases} \dfrac{t\,e^{-.2t}}{25} & \text{if } t > 0 \\ 0 & \text{otherwise} \end{cases}$$

6.23 This is an excellent problem, and it will be well worthwhile for you to work it two different ways. One way is to work out the density function for the random variable $Z = \max\{X, Y\}$, and then to compute $E(Z)$ using this density function. The second way is to simply utilize the joint density, since Z is already expressed as a function of X and Y. Since this requires figuring out how to integrate the function of two variables $g(x, y) = \max\{x, y\}$, the picture below should be helpful.

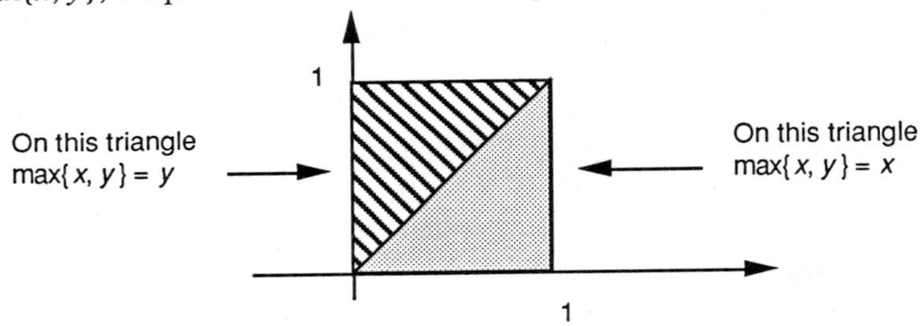

6.24 The expected value is 7/3.

Chapter 7

7.3 Denote by Y the number of the trial on which the second success occurs. In order for the second success to occur on trial n, there must be exactly one success during the first $n-1$ trials. Given any two *specific* trials among the first n trials, the probability of success occurring on those 2 and failure on all others would be p^2q^{n-2}. How many ways are there in which the second success could occur on the nth trial? The first must then occur on one of the trials from trial 1 to trial $n-1$, so there are $n-1$ ways for the second to occur on the nth trial. Thus

$$P(Y = n) = (n-1)\,p^2q^{n-2}$$

Another way of considering this problem is to think of X being the number of the trial on which the first success occurs, in which case

$$P(Y = n) = P(X = 1)\, P(Y = n \mid X = 1) + \cdots + P(X = n-1)\, P(Y = n \mid X = n-1)$$

Each term in the sum on the right side of this equation can be checked to be equal to p^2q^{n-2}, and there are $n - 1$ terms.

7.5 If Y_1 and Y_2 are the times of the first and second occurrences respectively, then $Y_2 = Y_1 + (Y_2 - Y_1)$. This simply amounts to looking at Y_2 as the time to the first occurrence plus the time between the first and second occurrence. From part 3 of Proposition 7.3, Y_1 and $Y_2 - Y_1$ are exponential with parameter μ = intensity of Poisson process, and they are independent. So Y_2 is the sum of two independent exponential random variables with parameter μ. This type of problem was treated in Chapter 6. See for example Problem 6.27.

7.6 See the comment above regarding Problem 7.5.

7.7 Let X denote the number of breakdowns to occur during 6 months. Then X is Poisson with parameter $\lambda = \mu t$ = (10 per year) × (.5 years) = 5. And the probability of 7 or more occurring is then

$$P(X > 7) = 1 - \{ P(X = 0) + \cdots + P(X = 6) \}$$

7.8 X is Poisson with parameter $\lambda = \mu t$ = (30 per hour) × (1/6 hour) = 5. And Y has the same distribution. $X + Y$ is Poisson with parameter

$$\lambda = \mu t = 30 \text{ per hour} \times \frac{1}{3} \text{ hour} = 10$$

Thus, using the properties of a Poisson process described in this section, there is no work to do in this problem. However, there is a basic underlying fact at work: If X is Poisson with parameter λ_1 and Y is Poisson with parameter λ_2 and X and Y are independent, then $X + Y$ is Poisson with parameter $\lambda_1 + \lambda_2$.

Chapter 8

8.3 First work out a reliability expression for the circuit. The following is one possibility:

$$\text{rel} = [\ (p_B + p_D - p_B p_D)\, p_C\ + p_A - p_A p_C\ (p_B\ + p_C\ - p_B p_D)\,]\, p_E$$

Now since each of the components is assumed to have exponentially distributed lifetime with parameter 1, this means that for any given device, at time t the probability that the device is still functioning is e^{-t}. So for every one of the probabilities on the right side of this equation, we can substitute e^{-t}. This gives the reliability of the system as a function of t, and works out to be

$$\text{rel}_{\text{sys}}(t) = e^{-2t} + 2e^{-3t} - 3e^{-4t} + e^{-5t}$$

Let's denote by X the time to failure for the whole circuit. So $(X > t)$ is the event that the circuit is still working at time t, and this is what $\text{rel}_{\text{sys}}(t)$ denotes above. Therefore

$$F_X(t) = P(X \le t) = 1 - \text{rel}_{\text{sys}}(t) = 1 - \{\ e^{-2t} + 2e^{-3t} - 3e^{-4t} + e^{-5t}\ \}$$

Differentiation gives the density function for X, and it is then easy to compute $E(X)$, which is the mttf for the circuit. Is it clear to you that F_X is a density function? If not, you should check out that it is. Can you make that observation based on the fact that $1 + 2 - 3 + 1 = 1$?

8.5 $f(t) \ = \ .3e^{-.3t} + .05e^{-.05t} - .35e^{-.35t}$

8.7 (a) $\mu/(\lambda + \mu) = 30/31 \approx .968$; that is, 96.8% of the time the equipment is operable.

(b) $f(5) \ = \ \left(1 - \dfrac{\mu}{\lambda + \mu}\right) e^{-5(\lambda+\mu)} \ + \ \dfrac{\mu}{\lambda + \mu} \ \approx \ .9735$

Appendix B: Values of the Normal Distribution Function

The table of values appearing here was generated on a Macintosh microcomputer using an elementary program written in Turbo™ Pascal to implement the algorithm shown in Problem 5.14 of this text. Values shown are for $\Phi(x)$, where Φ is the standard normal distribution function and $x > 0$. For $x < 0$, the identity $\Phi(-x) = 1 - \Phi(x)$ may be used.

x	0	1	2	3	4	5	6	7	8	9
0.00	0.5000	0.5040	0.5080	0.5120	0.5160	0.5199	0.5239	0.5279	0.5319	0.5359
0.10	0.5398	0.5438	0.5478	0.5517	0.5557	0.5596	0.5636	0.5675	0.5714	0.5753
0.20	0.5793	0.5832	0.5871	0.5910	0.5948	0.5987	0.6026	0.6064	0.6103	0.6141
0.30	0.6179	0.6217	0.6255	0.6293	0.6331	0.6368	0.6406	0.6443	0.6480	0.6517
0.40	0.6554	0.6591	0.6628	0.6664	0.6700	0.6736	0.6772	0.6808	0.6844	0.6879
0.50	0.6915	0.6950	0.6985	0.7019	0.7054	0.7088	0.7123	0.7157	0.7190	0.7224
0.60	0.7257	0.7291	0.7324	0.7357	0.7389	0.7422	0.7454	0.7486	0.7517	0.7549
0.70	0.7580	0.7611	0.7642	0.7673	0.7704	0.7734	0.7764	0.7794	0.7823	0.7852
0.80	0.7881	0.7910	0.7939	0.7967	0.7995	0.8023	0.8051	0.8078	0.8106	0.8133
0.90	0.8159	0.8186	0.8212	0.8238	0.8264	0.8289	0.8315	0.8340	0.8365	0.8389
1.00	0.8413	0.8438	0.8461	0.8485	0.8508	0.8531	0.8554	0.8577	0.8599	0.8621
1.10	0.8643	0.8665	0.8686	0.8708	0.8729	0.8749	0.8770	0.8790	0.8810	0.8830
1.20	0.8849	0.8869	0.8888	0.8907	0.8925	0.8944	0.8962	0.8980	0.8997	0.9015
1.30	0.9032	0.9049	0.9066	0.9082	0.9099	0.9115	0.9131	0.9147	0.9162	0.9177
1.40	0.9192	0.9207	0.9222	0.9236	0.9251	0.9265	0.9279	0.9292	0.9306	0.9319
1.50	0.9332	0.9345	0.9357	0.9370	0.9382	0.9394	0.9406	0.9418	0.9429	0.9441
1.60	0.9452	0.9463	0.9474	0.9484	0.9495	0.9505	0.9515	0.9525	0.9535	0.9545
1.70	0.9554	0.9564	0.9573	0.9582	0.9591	0.9599	0.9608	0.9616	0.9625	0.9633
1.80	0.9641	0.9649	0.9656	0.9664	0.9671	0.9678	0.9686	0.9693	0.9699	0.9706
1.90	0.9713	0.9719	0.9726	0.9732	0.9738	0.9744	0.9750	0.9756	0.9761	0.9767
2.00	0.9772	0.9778	0.9783	0.9788	0.9793	0.9798	0.9803	0.9808	0.9812	0.9817
2.10	0.9821	0.9826	0.9830	0.9834	0.9838	0.9842	0.9846	0.9850	0.9854	0.9857
2.20	0.9861	0.9864	0.9868	0.9871	0.9875	0.9878	0.9881	0.9884	0.9887	0.9890
2.30	0.9893	0.9896	0.9898	0.9901	0.9904	0.9906	0.9909	0.9911	0.9913	0.9916
2.40	0.9918	0.9920	0.9922	0.9925	0.9927	0.9929	0.9931	0.9932	0.9934	0.9936
2.50	0.9938	0.9940	0.9941	0.9943	0.9945	0.9946	0.9948	0.9949	0.9951	0.9952
2.60	0.9953	0.9955	0.9956	0.9957	0.9959	0.9960	0.9961	0.9962	0.9963	0.9964
2.70	0.9965	0.9966	0.9967	0.9968	0.9969	0.9970	0.9971	0.9972	0.9973	0.9974
2.80	0.9974	0.9975	0.9976	0.9977	0.9977	0.9978	0.9979	0.9979	0.9980	0.9981
2.90	0.9981	0.9982	0.9982	0.9983	0.9984	0.9984	0.9985	0.9985	0.9986	0.9986
3.00	0.9987	0.9987	0.9987	0.9988	0.9988	0.9989	0.9989	0.9989	0.9990	0.9990
3.10	0.9990	0.9991	0.9991	0.9991	0.9992	0.9992	0.9992	0.9992	0.9993	0.9993
3.20	0.9993	0.9993	0.9994	0.9994	0.9994	0.9994	0.9994	0.9995	0.9995	0.9995
3.30	0.9995	0.9995	0.9995	0.9996	0.9996	0.9996	0.9996	0.9996	0.9996	0.9997
3.40	0.9997	0.9997	0.9997	0.9997	0.9997	0.9997	0.9997	0.9997	0.9997	0.9998

Index

U

V

W